Marketing

A PRACTICAL GUIDE FOR
FISH FARMERS

Marketing

A PRACTICAL GUIDE FOR FISH FARMERS

SUSAN A. SHAW
Professor of Marketing,
University of Stirling

FOREWORD BY

MAURICE CANTLEY
Director, Marketing Development,
Highlands and Islands Development Board

PUBLISHED BY
FISHING NEWS BOOKS
IN ASSOCIATION WITH THE
HIGHLANDS AND ISLANDS
DEVELOPMENT BOARD

First published by Fishing News Books in association with the Highlands and Islands Development Board 1990

British Library
Cataloging in Publication Data
Shaw, Susan A.
Marketing: a practical guide for fish farmers
1. Great Britain. Fish farming
639.30941

ISBN 0–85238–176–X

Fishing News Books
A division of Blackwell Scientific Publications Ltd
Editorial Offices:
Osney Mead, Oxford OX2 0EL
(Orders: Tel. 0865 240201)
25 John Street, London WC1N 2BL
23 Ainslie Place, Edinburgh EH3 6AJ
3 Cambridge Center, Suite 208, Cambridge, MA 02142, USA
54 University Street, Carlton, Victoria 3053, Australia

Photoset by Enset (Photosetting), Midsomer Norton, Bath, Avon
Printed and bound in Great Britain by Hartnolls Ltd, Bodmin, Cornwall

Contents

Foreword

During recent years fish farming has been expanding rapidly in many countries around the world, and its progress in the Scottish highlands and islands has been particularly significant. Located almost entirely around the western and northern coastlines and mainly within the designated area of the Highlands and Islands Development Board (soon to become Highlands and Islands Enterprise), fish farming is very important to the economy of this area which accounts for over half of the land area of Scotland, but has only 7% of the population. The number of people employed directly in fish farming is already equivalent to more than 1700 full-time jobs and its value (at farm gate prices) to the local economy is more than £100 million and rising rapidly.

We recognise the importance of good marketing to the successful development of all industries and businesses. In the Highlands and Islands the majority of businesses are small and cannot support specialist marketing staff; many have been started by people with expertise in producing things rather than in selling them. We believe therefore that the Highlands and Islands Development Board and its successor Highlands and Islands Enterprise have an important role in assisting companies in our area to adopt a marketing approach to the development of their businesses. Marketing is much more than promotion or selling. It means looking at everything from the customer's point of view rather than from the point of view of what the company can or wants to produce.

Many fish farmers have had little or no previous experience of marketing so we thought that a guide which signposted the main issues would be useful. It had to be easy to read and address itself directly to the situation facing the small- to medium-sized fish farmer in Scotland in the 1990s: a practical, hands-on guide, not a theoretical textbook. We were therefore very pleased when Susan Shaw, Professor of Marketing at the University of Stirling agreed to write it for us. Her extensive knowledge of marketing linked to her valuable work on aquaculture made her the ideal choice.

Experience shows that fish farmers throughout the world face similar marketing situations, and although this book was written for Scottish fish farmers, it should be as relevant to fish farmers in other countries as it is to those in Scotland. While most of the examples pertain to the Scottish scene, they should translate to whatever country you operate in.

We hope that this book will be of interest and practical value to those involved in all types of fish farming - shellfish, salmon, trout or other finfish species – particularly in the Scottish Highlands and Islands but also further afield.

Inverness July 1990

Acknowledgements

I am indebted to many people connected with the Scottish fish farming industry for help but I would particularly like to thank Kevin Gruer and Maurice Cantley of the Highlands and Islands Development Board for the support and advice which they have given me at all stages of this project. I am grateful to William Crowe, Scottish Salmon Growers Association; Dennis Overton, Aquascot Marketing Limited and Gitte Salvarli, Association of Scottish Shellfish Growers for reading and making valuable comments on earlier drafts of the book. Alan Munro of the Department of Agriculture and Fisheries for Scotland gave valuable advice on health regulations, as did Iain Goldie of LEP Perishables on export regulations. Any errors, of course, remain the responsibility of the author alone.

1
The Marketing Approach to Business

What is marketing?
The scope of marketing
Market driven businesses

WHAT IS MARKETING?

There are many definitions of marketing but my own favourite because it is short is:

> 'The primary business of every business is to stay in business and to do that you have to get and keep customers.' (Levitt)

To be profitable a business has to find out what customers want and make sure that this is what they are offered. Customers always have a right not to buy and if you do not offer them what they want, there are likely to be competitors around who will. It is worth asking two questions:

- Why should customers buy from me rather than from a competitor? What can I offer in terms of products or levels of service that makes customers happier to do business with me rather than with someone else?
- Why should people buy shellfish, salmon or trout when they could choose around 6000 other items in the typical supermarket?

The typical eater of shellfish and finfish is affluent and demanding, and because of the wide range of high quality alternatives on offer they will readily switch if not satisfied. Research evidence shows that it is the businesses which understand and respond to this which are the most successful.

THE SCOPE OF MARKETING

Good marketing is far more than just finding customers for what farms have decided to produce. It starts with customers - both the final end user and trade customers – and from an analysis of their needs works back to decisions about what should be produced, when it should be produced and what the customer should be offered. It is a three stage process (see Fig. 1.1).

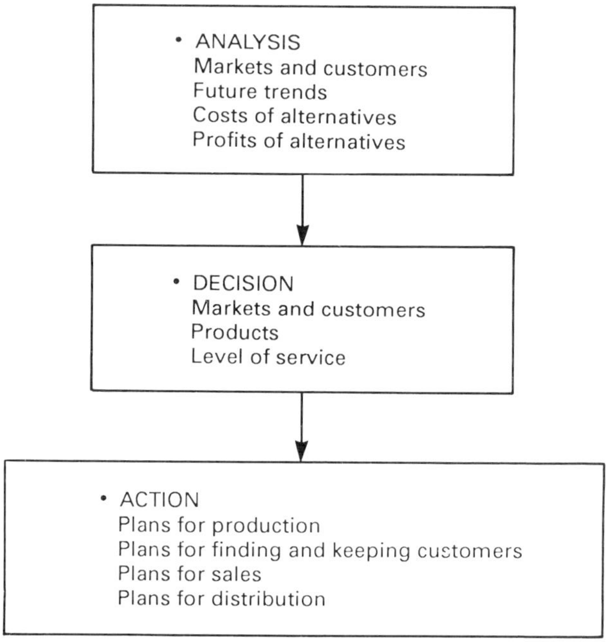

Fig. 1.1

MARKET DRIVEN BUSINESSES

Most of this book is about managing marketing but when reading it, don't forget that this is only half the story. Marketing thinking should not be separated from the rest of the business and in particular the business must be completely committed to the need for customer satisfaction, down to the smallest detail.
Examples of how *not* to do it are:

- Planning production before planning marketing.
- Scheduling harvesting for the convenience of the farm, not the customer.
- Producing the sizes of fish or shellfish that are easy, but not the ones customers want.
- Getting irritated when customers are fussy about colour, size and harvesting methods.
- Producing beautiful fish or shellfish but handling them badly.
- Ignoring customer specifications because they are a nuisance.

The marketing approach is particularly vital for fish farming because the industry is so competitive. As supplies increase customers are becoming less and less tolerant of deficiencies in quality and service. It is becoming more and more necessary for each business to have a competitive edge over its rivals to succeed. Marketing is about creating this competitive edge.

CHECKLIST

Do you understand your customers? Do you know why they buy from you?

....

Do you listen to your customers – regularly?

....

Give examples of how responsive your business is to your customers.

....

Give examples where you have gone out of your way to help your customers over the last year.

....

Are all your decisions in all parts of the business driven by considerations of your customers?

....

2
What Kind of Business?

Where are we going?
The choices
Making the links
Lessons from the market
The business size factor
The future for small fish farmers
Adding value
Taking decisions
Managing marketing

WHERE ARE WE GOING?

Any business must have a view of where it is going and what it wants to be. Without this, it is difficult to make the right decisions and the business will drift along, with the danger that it will miss opportunities or ignore problems until it is too late to tackle them effectively.

THE CHOICES

The range of choices facing the small-scale fish or shellfish farmer include whether to:

- Sell fresh unprocessed product?
- Sell one product or a bigger range?
- Add value by processing products?
- Choose local customers, sell to more distant customers or a combination of both?
- Export?
- Do your own marketing or get other people to do it for you?

Decisions will have a fundamental effect on the business for a considerable time into the future. This is because these choices commit you to:

- A heavy investment of time and effort in building up a strong and loyal customer base. It is not enough to get trial orders. Customers have to be convinced to place regular orders and to stay with you through time.
- A major investment of time and effort in becoming the type of organisation that can satisfy customers effectively.

It is worth taking time at the start to make sure that you choose the right option.

MAKING THE LINKS

Different options have different profit potential. The key to success is to choose the option which fits best in terms of:

- The opportunity,
- the costs of meeting the opportunity,
- your ability to compete effectively.

Opportunities, costs and competitiveness all go together – they cannot be looked at separately. Choosing which way to go is like making sure that the links of a chain all join together (see Fig. 2.1).

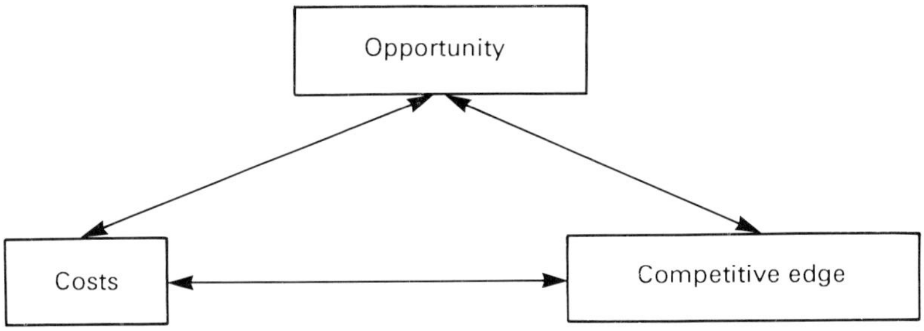

Fig. 2.1

If one of the links is missing, the chain will break. Two points must be remembered:

- There may be opportunities in some markets, but the costs of selling to them make them unprofitable. There may be opportunities to supply some customers, but it may be better to choose other customers who can be supplied more profitably. It is not enough to see an opportunity, it must have *profit potential.*
- Not all businesses are the same. The right opportunity for one company may be the wrong one for another. The key is to choose the *right* opportunity for *your* business.

LESSONS FROM THE MARKET

Some other points to bear in mind are:

- *Competition*
  There may be a few specialist markets but on the whole customer requirements tend to be similar. In any option you choose, you cannot escape the need for a given quality and to be competitive on price.
- *Customer demand*
  There are exceptions, but most trade customers require product regularity and are looking for a supplier or group of suppliers who can deliver this.
- *Rising expectations of customers*
  With increasing supplies, more and more trade customers are becoming demanding in the level and number of their requirements, and less tolerant when these are not met. This used to be associated with the more demanding accounts such as multiple retailers but it is now spreading to other trade customers.

THE BUSINESS SIZE FACTOR

This points to the increasing importance of size in marketing. Larger businesses have advantages:

- *Customer benefits*
  Regular deliveries of standard product;
  Supplying larger orders;
  More careful grading and selection because of larger volumes.
- *Cost savings*
  Economies in marketing overheads;
  Lower transport costs because of larger, more frequent loads.
- *Competitiveness*
  Able to supply a wider range of customers and spread risks;
  Able to employ specialist staff who will:
  raise the quality of marketing,
  tackle more difficult markets especially export markets;
  Able to raise capital for expansion more easily;
  Operate at a scale in processing which is competitive and which can lead to expansion into value-added products.

Studies have suggested that there are penalties in marketing less than 1500 tonnes of finfish or £400,000 of shellfish annually, and that marketing and other costs continue to fall per tonne of output until the business is handling at least 3000 tonnes of finfish or over £1 million of shellfish.

THE FUTURE FOR SMALL FISH FARMERS

Does this mean that there is no future in being small? No – but it does mean that small fish farmers must organise themselves so that they can concentrate on their *competitive advantage* – the things that they are good at.

Small fish farmers are good at production. Many industry experts believe that the small owner-managed fish farm is one of the most cost effective ways of growing good quality finfish and shellfish because owner-managers take the greatest care. Smaller fish farms can, where opportunities exist, also build up their businesses by selling to local customers such as hotels and shops and by direct sales from farm shops. Here the personal touch and high level of local service give the business its competitive edge.

Smaller businesses face difficulties when their production is higher than can be sold locally and they have to compete more directly with larger businesses. There are ways in which this can be done but it involves changing to a different type of organisation (see Fig. 2.2).

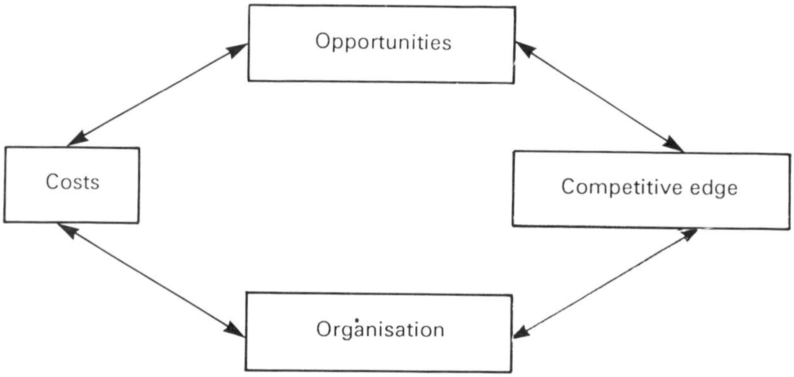

Fig. 2.2

Options are:

- Joint marketing with other fish farmers;
- Contract on-growing for a large producer.

Chapter 3 discusses the pros and cons of these alternatives. They are ways in which the advantages that you enjoy in production can be linked to strong and effective marketing.

ADDING VALUE

One option is for fish farmers to add value to products by filleting, steaking, smoking, marinading, etc. Adding value makes a good deal of sense – for the industry as a whole and for the larger businesses because of:

- Opportunities for additional profit;
- Opportunities for extending sales to new customers;
- Reduced risks from a wider product range.

A good example of success is the US catfish industry where producers combined to form a co-operative. In addition to fresh sales, the co-operative opened a processing plant and has successfully marketed a range of catfish products throughout the United States. By extending markets, the over-supply situation was reversed so that catfish farmers now receive higher prices for their fish.

However, whether it makes sense for a small fish farmer to do this alone is an entirely different matter. It *may* make sense to move into basic processing activities such as filleting and gutting if:

- Existing customers want this extra service;
- You can use contract facilities or have access to facilities which allow you to be competitive on costs.

In general the further you try to go into added value, the more problems you will meet:

- *New markets*
  New markets often need new skills with different customers and different ways of doing business.
- *Standards*
  Standards in food processing are rising and the costs of investing in plant and equipment to meet standards are considerable.
- *Competition*
  Added value is not necessarily added profit. There are many companies already in the value-added fish products business. Unless you have a very specialist product your costs will be higher than those of businesses operating on a larger scale.

There are exceptions – there are small Scottish companies who have succeeded despite these problems. When you look at them carefully they all have one thing in common. They have special skills such as excellent marketing abilities or very good and unique products that have enabled them to build up specialist markets. Not everyone has these skills. Just because others have done it, don't rush in. Ask yourselves:

- Do we have something unique that offsets the marketing problems of being small?

If the answer is *no* . . . *KEEP OUT!*

TAKING DECISIONS

If you join a marketing group or get involved in contracts, the processes of analysis, decision-making and action described in Chapter 1 will be led by your marketing advisers. If you work alone, this will be your own responsibility. In each case the questions to be answered are the same.

THE OPPORTUNITY

How big is this market?

....

Will it grow?

....

How easy is it to get in?

....

How stable is it?

....

How price conscious is it?

....

What is the trend in prices? How much do prices fluctuate?

....

What is needed in terms of:

Products

(1) Sizes ....

(2) Quality ....

(3) Volumes ....

(4) Timing ....

Packaging ....

Payment ....

Customer Contact ....

Delivery ....

COMPETITION

Who would be our main competitors? What others are there?

....

How strong are the main competitors?

Products ....

Organisation ....

How do we compare with them? What are our advantages and disadvantages?

....

....

PROFIT

If the answers to any of these questions are unfavourable to you, you will have to think carefully about whether the option is the right one.

If the opportunity still looks good, the profit which you might make must be calculated by comparing the money that you would receive with *all* the costs of supplying the product:

Profit = Price × Volume of sales − Costs of
product,
processing,
packaging,
administration,
selling,
transport,
overheads (lighting, heating, etc.),
interest,
trade association subscriptions,
marketing levies.

Don't forget to cost in your own time and that of your family in the calculations!

It is also useful to see how risky the opportunity is. For instance:

Would you still make money if prices fell by 10%, 20%?
Would you still make money if costs rose by 10%, 20%?
Would you still make money if interest rates rose by 2%, 3%?

If it looks as if the operation could easily be in danger of running at a loss if conditions change, this is a warning not to go ahead.

MANAGING MARKETING

If things still look good at this stage, make a plan for carrying out your ideas. This will help to control activities by giving a clear sense of purpose and direction. The plan should be quite short but it should contain all the things listed below.

MARKETING PLAN

What we aim to achieve

....

Value of sales

....

Return on investment

....

Share of market

....

How we are going to do it

Market(s)

....

Customer(s)

....

Sales plan

....

Distribution plan

....

Marketing Plan (cont.)

Promotion and selling

....

A timetable:

Activity	*Target date*
....	....
....	....
....	....
....	....

Budgets
Targets for income and costs

	£
Sales	....
Costs	....
products	....
processing	....
packaging	....
administration	....
selling	....
transport	....
overheads lighting	....
heating	....
other	....
interest	....
trade association subscriptions	....
marketing levies	....
Total	....
Balance: planned profit	....

Check from time to time that you are keeping to your plan. Try to explain differences between your forecasts and actual events – this will lead to a better understanding of your markets and how they work.

3
Joint Marketing

MARKETING GROUPS – THE OPTIONS

Chapter 2 showed that the benefits of marketing with others are:

- Being a member of a large marketing company able to compete on equal terms;
- The freedom to leave marketing to the specialists and concentrate on what you are good at and probably enjoy most – production.

There are a number of ways in which you can join together:

- A marketing company set up by a group of farmers;
- A marketing company set up jointly by a trading company and a group of growers;
- A marketing co-operative of farmers.

MARKETING COMPANIES

Ownership	Limited liability company owned by the farmers as shareholders. Structure of shares subject to agreement. Votes in proportion to shares held. Shares not necessarily in proportion to tonnages sold.

Marketing Companies (cont.)

	Can admit new members but only with agreement of majority of shareholders. Provision for members to leave the company if dissatisfied.
Sources of finance	Share capital of members, bank loans, grants.
Profits	Dividends on shares.
Law	Companies Act 1985.
Control	Directors elected by shareholders. Works best with 5–12 people to give range of skills, fair representation of members but not top heavy. Members of management team can be elected as executive directors.
Management	Company run by management team of: marketing director, technical controller, sales assistants, administrative staff. Responsible to Board of Directors.

FARMER–TRADING COMPANY MARKETING GROUP

Ownership	Limited liability company of farmers and trading company. Structure of share ownership subject to agreement but trading company likely to require substantial shares for marketing control.
Sources of finance	Shares, loans, grants.
Profits	Dividends on shares.
Law	Companies Act 1985.
Control	Directors elected by shareholders – as above. Normally some provision for heavy representation of trading company.

MARKETING CO-OPERATIVE	
Ownership	Membership is through nominal shareholding and is normally open to all businesses engaged in the relevant activity e.g. shellfish farming, salmon ongrowing.
Sources of finance	Shares, member investment (but interest allowed small), grants, loans.
Profit	Main returns to members through prices achieved but prices have to allow for profits for reinvestment and to secure the financial base. Guidance and draft rules available from Scottish Agricultural Organisation Society (SAOS).
Control	Voting rights are approximately equal for all members and are not in proportion to sales through the co-op or money lent. Members elect a Board of Directors of 5–12 people.
Management	As for marketing groups.

There are some differences between the three types of marketing company:

Marketing group	Shares help to ensure commitment. Dividends help to give a marketing orientation because they give members a direct interest in the profits of the marketing company.
Marketing group: Farmers–Trader	The growers gain from the marketing experience of the trading company. A means of starting a business quickly by using the experience of the trader.
Marketing co-operatives	The most democratic structure but has not been popular in fish farming because there is no direct return on money invested. There can be problems because of over-emphasis on getting short-term high prices for members at the expense of long-term development, nevertheless there are successful examples.

If you want more details about setting up marketing companies, contact the major banks or the Highlands and Islands Development Board (HIDB)/Highlands and Islands Enterprise (HIE).

All these routes are feasible but probably marketing groups are the easiest to run in a market orientated way. This is because of the emphasis they give to profit from marketing activities as well as production. Whether the group joins with a trading company really depends on whether there are suitable partners available who are prepared to commit themselves to the company and not just regard it as a sideline.

HOW TO SUCCEED – ORGANISATION, RULES AND ATTITUDES

Three elements are required to make groups successful:

- An organisation that everyone is happy with;
- Sensible rules;
- Mutual commitment and trust.

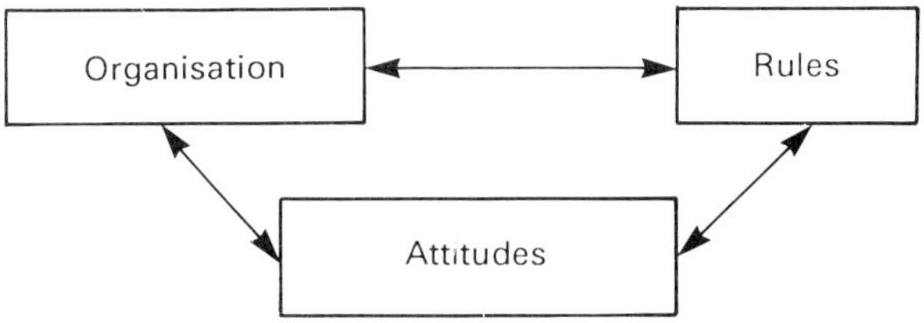

Fig. 3.1

Trust and commitment are the most necessary of the three elements. Members have to accept that belonging to a group is different because it means:

- Giving up independence;
- Working to the aims of the group, not the individual.

Ways in which members have to give up independence include:

- Producing to the output schedules of the group so that the group can sell year round in line with customer demand.
- Being flexible and willing to change harvesting schedules when the marketing company asks you to do so because of changes in market conditions.
- Producing to common quality standards.
- Accepting lower short-term prices so that the group can invest in market development for the future.
- Accepting majority decisions and allowing the management freedom in day to day action.

This is a lot to give up – but the benefits in return are considerable.

SETTING UP THE GROUP

The basic guide-lines are similar whichever type of organisation you choose.

The early stages are the most critical and it is at this time that the group has to make sure that members are willing to work to common objectives and to spend time in developing the group. Ask:

- Do the farmers who want to join have similar interests and thus are more likely to agree with each other about:
  size?
  costs?
  qualities?
- Is everybody willing to work to common quality standards?
- Are the logistics of joint marketing sensible and feasible in terms of:
  location of members?
  transport available?
- Are members really willing to invest their *time* as well as money in getting the group off the ground?

A business plan must be drawn up before the group is set up and grant aid is available in many cases to do this. This is a good opportunity to assess carefully the gains from the group and an ideal time for members to assess the commitment which they are taking on.

HOW THE GROUP WORKS

You should seek professional advice from lawyers or HIDB/HIE about the specific legal requirements, but there are some general guide-lines.

The board

The functions of the managing board must be clearly laid out and the board should be responsible for:

- Defining the objectives of the group;
- Approving policies;
- Setting up efficient management systems;
- Appointing a manager and monitoring his/her performance.

The following should be directly represented on the board:

- Marketing,
- Finance,
- Quality control,
- Processing (if relevant).

A well run board:

- Monitors the progress of the business regularly – say monthly;
- Sees that members are kept well informed;

- Does not interfere in the day to day running of the business unless it feels it is necessary;
- Does not interfere with the responsibilities of the managers if the job is being done properly;
- Listens to advice from the marketing experts.

The last point is very important. Don't lose sight of the marketing objectives of the group. Farmers must respond to the needs of the marketing company and not the other way round, so you must listen to what the marketing experts are saying. Don't impose your views on them.

Management

The management team is responsible for:

- Carrying out day to day business efficiently and profitably;
- Designing policies;
- Running the management system.

This means

- Preparing and implementing marketing plans and reporting on progress;
- Market research and market monitoring;
- Co-ordinating production to meet marketing plans;
- Organising processing (where relevant);
- Buying supplies for members (where relevant);
- Representing the company.

It is crucial for the group that this is done well. This means that the group must employ highly professional staff with relevant experience. Since there is a shortage of good marketing people in the food industry, this means that you must offer high and competitive salaries.

Supply agreements

The group must operate as a single marketing unit and not a collection of separate companies. This means:

- Production schedules worked out by the management team and kept to as far as possible by members;
- Committing a high percentage of output to the group i.e at least 90% and preferably 100%. This is *essential* to ensure that the group can count on your supplies. It ensures that you take your commitment seriously. It avoids undercutting of group price or the group being used simply as a clearing house for lower quality produce.
- Quality control. The marketing group must have the power to dictate standards to members and reject unsatisfactory products. It must be able to develop consistent product across farms, otherwise it cannot act as a single company.

Prices and payment

The best system is one where the marketing company buys fish or shellfish from members and combines output from different farms using the marketing group's name for marketing.

The fixing of prices can be a difficult issue because farmers naturally want the highest prices they can get and to be able to sell at the time of year when prices are at their peak. At the same time, if the prices offered to members are so high that they starve the marketing company of money it will not be able to develop the business of the group.

The best system is one where farmers are paid:

Average monthly or quarterly prices minus An agreed levy to finance the marketing company

In this way:

- Conflict is avoided because there is a simple rule that everyone understands.
- Price fluctuations during the period are evened out.
- Because everyone gets the same price it reinforces the need to have the same quality standards.
- The marketing company has an incentive to make as much profit as possible, because the higher the prices they sell at the more cash they will generate.

FOLLOWING SUCCESS

It takes considerable effort to set up a good group, but you may not need to do this. There are already marketing groups for farmed fish in Scotland and it might be sensible, as a first option, to consider whether you could join them. This would have the advantage of avoiding the effort involved in setting up. A second general advantage for fish farmers is more effective marketing if, rather than lots of small groups, there are a small number of large groups with all the benefits of size. This applies equally to shellfish as well as to salmon and trout.

CONTRACTS

The issues

The other way of turning small size to advantage, rather than disadvantage, is to produce for other people under contract. This can either be for another, larger fish farmer or for a trading company. In comparison with the marketing company this option is a balance between

EFFORT	and	RISK

The benefits are:

- Marketing responsibility is transferred to the other company;
- Possible access to the quality control and technical advice of the bigger company;
- Savings in transport costs (depending on location);
- Less involvement and responsibility for marketing than with marketing groups.

These all help the farmer to concentrate on production. But there are risks in going this route:

- Dependence on one customer, but with very little control over him;
- The difficulty of finding other customers if the contract is ended, especially when there is a lot of fish around;
- The danger that the marketing of your output will be treated as less important by your partner.

These can be problems but this may still be a very worthwhile option to consider if you find attractive partners who are good at marketing and well organised. In any case there are ways in which you can minimise risks with the right contract.

GUIDE-LINES FOR CONTRACTS

- A long term commitment, i.e. after an initial trial period, a minimum of 3–4 years with mutually agreed termination conditions.
- A commitment by your partner to market your output with his other sales, not as a separate exercise.
- Agreed quality control standards and monitoring procedures.
- Pricing system based on prices achieved with agreed deduction for marketing costs.
- Find other fish farmers who wish to operate in this way, so that you combine together to contract with your partner. The more output you have to control and offer, the more power you will have to ensure that the terms of the agreement are fair.

FOUR QUESTIONS TO ANSWER WHEN GOING THE JOINT MARKETING ROUTE

1 What are the benefits to your business?

....

2 Have you measured the benefits – has a feasibility study been carried out?

....

3 What challenges face you and how are you going to tackle them?

Challenge Action Plan

....

....

....

4 How are you ensuring that the marketing will be done in a customer orientated way?

....

REFERENCE

Salmon Operational Guide-lines Study by Fintrac, for the Highlands and Islands Development Board and the Scottish Salmon Growers Association, available from the HIDB/HIE Product Marketing Division.

USEFUL ADDRESSES

Scottish Agricultural Organisation Society (SAOS)
18 Claremont Crescent
EDINBURGH EH7 4JW

Food From Britain
301–344 Market Towers
New Covent Garden Market
LONDON SW8 5NQ

Highlands and Islands Development Board (HIDB)/Highlands and Islands Enterprise (HIE)
Bridge House
20 Bridge Street
INVERNESS IV1 1QR

Also see Useful Addresses section at the end of the book for further information.

4
How Markets Work

INTRODUCTION

The most common explanation of failures or poor performance, particularly for small businesses, is the impact of some outside uncontrollable force for which the business has not properly prepared. There could be sudden price reductions, or major customers could go out of business. Rising costs could force you to seek higher prices in order to survive, but from customers who are reluctant to pay.

Often these changes are out of your control. However, if you are aware of the forces at work, you will be in a much better position to anticipate them, to respond in the most effective way and sometimes to use them to your advantage.

MARKET FORCES

Market conditions are never stable – particularly in markets for food products like fish. This is reflected in prices which can change from year to year, month to month and even day to day. These changes have many causes but they can be grouped into those coming from customers (*demand*) and those caused by producers themselves (*supply*). Changes from either side put pressure on prices to change (see Fig. 4.1).

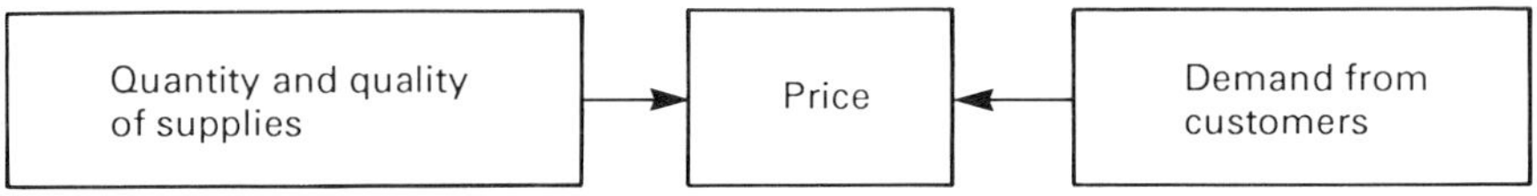

Fig. 4.1

If customers want to buy less but supplies stay at the same level or rise, prices fall because producers compete with each other to maintain their sales levels. In this situation customers are in a strong position to dictate prices and to put pressure on producers to lower them. If customers want to buy more but supplies are not increasing, the reverse happens. Customers now have to compete with each other to get supplies and producers are in a strong position to ask for higher prices (see Fig. 4.2).

Situation		Prices
Demand	*Supply*	
Growing	Constant	↑
Growing	Falling	↑
Static	Increasing	↓
Falling	Constant	↓

Fig. 4.2

DEMAND

DEMAND CHANGES BECAUSE OF CHANGES IN:

- Prices of other fish.
- Prices of other competing foods.
- Marketing and consumer choice.
- Consumer spending power.

- *Prices of other fish*
  Rising or falling prices for other types of fish have an effect. Prices of farmed salmon in North America during the summer months are pushed down by the large harvests of wild salmon. 1–2 kg trout and salmon are strong competitors with each other. Falling salmon prices

bring the prices of trout in this size range down with them. The price of queen scallops is affected by prices of king scallops.

- *Food prices*
  Food competition is complex. Mussels for instance have many competitors. People may choose other fish, melon, paté or soup instead of mussels, and relative prices influence this choice. This can work in either direction. If the prices of mussels fall, this will help to create new demand. If the prices of other competing products fall, the effect will be the reverse.

- *Marketing and consumer choice*
  People take many factors into account when deciding what to buy:
  taste,
  colour appeal,
  ease of preparation and cooking,
  nutrition value,
  packaging,
  shops and what they have available,
  the season.

- *Consumer spending power*
  In richer countries people do not eat more food when they have more money to spend. However, they are likely to spend more on highly priced foods and higher quality foods. One of the reasons why demand for farmed fish and farmed fish products has grown is rising incomes of people in the main markets, especially Europe and North America. When people are less well off, perhaps because of rising mortgage rates or perhaps because incomes are rising less rapidly than inflation, demand will fall.

Producers themselves have some influence here. There are many ways in which good marketing can help to create demand.

- Product quality,
- Product packaging,
- Availability/distribution,
- Information,
- Persuasion.

THE SUPPLY PROBLEM

Demand tends to change more slowly than supply. Probably the main reasons for changes in prices are the decisions of producers.

Uncontrollable changes in weather conditions and water temperatures affect output and there is not a great deal to be done about this. Disease problems can be helped by careful husbandry but control is not perfect and output can be affected.

The main reason for supply changes lies with the production decisions of farms. When conditions are good and prices are high there is a natural tendency for farmers to want to expand. However, this assumes that the market can be stimulated to grow to absorb all the extra supplies. If it does, prices will continue to remain good and everyone has made the right decision. If it does not, prices will start to fall because this may be the only way in which people can be persuaded to buy more.

On the whole, producers tend to be over-optimistic and expand far too quickly, with not enough marketing effort to create new and bigger markets. This explains the problems of the trout industry in the early 1980s and salmon in the late 1980s. Eventually the problem will be solved, because supplies will fall back again. Some producers will decide not to re-stock at the same levels and others will go out of business altogether but this may take considerable time.

MANAGING SUPPLIES

Does this always have to happen? The answer is no – but sensible management of supplies is easier in some situations than in others:

- *Unstable markets*
  It is especially difficult to predict demand in new markets when demand is growing very rapidly. No one really knows for how long this will go on but it cannot go on for ever. It is best to be cautious.
- *Producer co-ordination*
  The smaller the number of sales outlets the more each will see the impact of their activities on the industry as a whole and the more sensible their production decisions are likely to be.

The British trout industry is a very good example of how producer co-ordination, through the formation of bigger companies and big co-operatives has reduced the number of selling points, and planned output increases cautiously. In the process the industry has been turned from one of low profit and low prices in 1980 to a successful and profitable industry in the 1990s. Similar opportunities are now available for shellfish farmers for whom larger groups will help to both improve marketing and create greater stability.

EXCHANGE RATES

The fish business is an international one – this introduces exchange rates which are another cause of instability into markets. Exchange rates between currencies are changing all the time:

- When the value of the currency of a customer country falls (e.g. the value of the US$ relative to £), customers in that country will buy less because in terms of their own currency the product has become dearer.
- When the value of their currency rises they will buy more because imports have become cheaper to them.

This does not only affect the export trade. If the currency of a major customer – say the United States – changes this will have effects in all markets because they have such a large share of the world total demand:

Decline in value of US$→decline in demand from US→falling prices in US→supplies diverted to European markets→falling prices in Europe.

Changing currency values affect competitors:

Decline in value of £→cheaper Scottish oysters in international markets →customers switch to Scottish oysters.

The effects will be greater the bigger the competitor. For instance, changes in the value of French francs will have a bigger effect on international oyster prices because they are large suppliers. Changes in the value of the Belgian franc will by contrast have a very small effect.

EXPLAINING PARADOXES

Sometimes the impacts of supply and demand changes appear confusing on the surface but there is always a logical explanation:

CHANGE		EXPLANATION
No premium for quality	(1)	Does the market want higher quality?
	(2)	Is there a temporary increase in supplies of higher quality and customers cannot be found quickly?
No premium for larger sizes	(1)	No demand.
	(2)	Temporary increase in supplies but customers cannot take more.
Different prices from different customers	(1)	Are they buying the same quality?
	(2)	Is the consignment the same size?
	(3)	Are they getting better service?

Explaining paradoxes (cont.)

Different producers get different prices	(1) Are they supplying the same quality? (2) Are they offering the same levels of service? (3) Are they supplying the same customers? (4) Are the prices at the same level? (Don't confuse 'ex-farm' with 'delivered' prices!)
Falling prices for producers but no change in retail prices.	(1) Bargaining power of producers is weak when markets over supplied, so wholesalers or retailers force price reductions but don't pass them on. (2) Need to stimulate consumer demand.

ACTION POINTS

Because you cannot escape the market, the only alternative is to anticipate and to organise.

- *Predict*

  You will be a better predictor of change if you understand what is happening now and what are the trends for the future.
- *Listen*

  Trade Associations spend time and money analysing markets. Listen to the advice which they give.
- *Organise*

  Do you have the resources to:

  analyse and predict markets?

  achieve bargaining power against other members of the trade?

  plan supplies in line with market conditions, not just harvesting needs?

  If the answer is no – re-read Chapter 3 about marketing groups.

CHECKLIST

How do you expect conditions to change over the next year:

Improvement ....

No change ....

Deterioration ....

List the reasons

1 ....

2 ....

3 ....

4 ....

How do you expect conditions to change over the next three years?

Improvement ....

No change ....

Deterioration ....

List the reasons

1 ....

2 ....

3 ....

4 ....

What actions do you plan to cope with these changes?

1 ....

2 ....

3 ....

4 ....

5
Keeping Up To Date With Markets

WHY COLLECT INFORMATION?

By using knowledge the company can make more effective decisions. Knowledge is needed to:

- Understand customer needs,
- Choose between different opportunities,
- Understand changes in markets.

DESIGNING THE SYSTEM

The first step is to decide what information is needed:

- What types of decisions do you have to make regularly?
- What types of information do you need for them?
- What information do you get now?
- What is the gap between the information you have now and what is wanted?
- Is the information available?
- Can you afford to collect it?

Collecting information for its own sake is a waste of time. The key to success is to collect only what is of value. Information can also be expensive, both in terms of time and money. Make sure that the investment in information is worthwhile by checking that you need it *and* that you make use of it. Throw away information which has become out of date.

TYPES OF INFORMATION

Information sources can be roughly divided into:

- *Internal*
  collected from inside the business,
- *External*
  collected from publications and individuals outside the business.

A mixture of both will be needed.

INTERNAL INFORMATION – MAKING THE MOST OF WHAT YOU ALREADY HAVE

Customer records

It is surprising how much information most companies already have, if it is collected and used properly. This information also has the advantage that it is inexpensive to collect.

Some of the most useful information comes from customer records. These can be used to:

- *See trends*
  Are orders increasing/falling?
  Are orders from different types of customers following different trends?
- *Help with planning*
  Is there a weekly ordering pattern?
  Are there seasonal patterns?
  Are you consistently getting better prices from some types of customers?
  Are you consistently getting larger orders from some customers?
  What prices are you getting for different qualities and at different times of the year?
- *Help with sales*
  Are some customers ordering more/less than before?
  Are there customer accounts which can be encouraged to grow?

Who are the non-active customers who have not ordered recently? Why is this? Can it be changed?

- *Negotiations*
  How sensitive are different customers to prices?
  What prices are being achieved with other customers?

This information is much more effective if it is collected in an organised way. A simple standardised form which can be used for all customers is a useful way of doing this (see Fig. 5.1).

CUSTOMER RECORD FORM

Customer name: ....

Address: ....

....

....

....

....

Telephone number: ....

Fax number: ....

Telex: ....

Contact name: ....

Week: .... Year: ....

ORDER

Product	Quality	kgs
....	....	....
....	....	....
....	....	....
....	....	....
....	....	....

Price specification: ....

Price: ....

Date of entry: ....

Fig. 5.1

USING THE ACCOUNTS

The order books and accounts of the business give valuable information which will help to:

- Manage customers' orders on a day to day basis.
- Identify trends to help longer term decisions.

So, it is useful to:

- Keep schedules of harvesting plans and orders, and look for patterns.
- Examine budgets and actual receipts and costs, and explain differences.
- Watch the costs of servicing different customers and markets.

EXTERNAL INFORMATION

Just talking:
Collecting market research data can be very expensive. Reports on markets and trends can be expensive to buy and you are unlikely to have the time to carry out surveys yourself. But there are ways in which any business can keep itself up to date in an affordable way by using the following:

- Customers
- Suppliers
- Distributors

The better your relationship with your customers, the more useful you should find them as a source of what is going on in the market because they will be more honest and open with you. Suppliers are often also well informed about your markets, especially when they are large international companies. There are other useful contacts – especially if you go to trade shows, exhibitions and conferences when people with similar interests to your own come together. These can be good sources of information for

- Trends,
- Views of the future,
- Information on customers,
- Information on competitors.

Like internal information, it is best to be systematic about this:

- *Report forms*
  Make a list in advance of the information you want.
  Write down interesting information on a report form (see Fig. 5.2).
- *Check reliability*
  How representative are the people you are talking to? Are they all saying the same things and if not, why is this – are they supplying different markets, have they different interests? Do you need to talk to more people to get a clearer picture? Do you need to talk to more people to check the accuracy of information?

CONTACT REPORT

Contact name: .... Report by: ....

Company name: ....

Date: ....

1. Price information

   Product: ....

   Date: ....

   Volumes: ....

2. Other market information

   ....

   ....

3. Competitor information

   ....

   ....

4. Other

   ....

   ....

Fig. 5.2

REPORTS AND NEWSPAPERS

There are many other sources of useful information on trends, customers and competitors. The seafood business is very fortunate in that there are a number of excellent trade newspapers (the main ones are listed near the end of the chapter) which contain articles written by experts on markets and marketing. Other useful sources are:

- Your trade association,
- Local Chamber of Commerce,
- Libraries, especially commercial sections of larger libraries,
- Sea Fish Industry Authority,
- HIDB/HIE.

ACTION POINTS

1. What types of market information do you need?

   ....

   ....

2. Is your internal marketing information up to date?

   ....

   ....

3. List the information available from your trade association.

   ....

   ....

4. When did you last discuss your information needs with your trade association?

   ....

   ....

5. Which useful information is held by which libraries?

   ....

   ....

6. What efforts will be made over the next year to talk to customers? Suppliers? Other experts on markets?

   Customers: ....

   Suppliers: ....

   Others: ....

MAGAZINES

Title	*Coverage*	*Details*
Fish Trader	Heavy bias to wet fish trade but covers wholesaling processing and retailing in UK. Weekly Billingsgate prices.	FMJ International Publications Ltd Queensway House 2 Queensway REDHILL, Surrey RH1 1QS *Tel:* 0737 768611 *Fax:* 0737 761685 Fortnightly
Seafood International	International journal of marketing of all types of seafood.	AGB Heighway Ltd Alexander House 81–89 Farringdon Road LONDON EC1M 3LL *Tel:* 071 405 0500 *Fax:* 071 405 8555 Monthly
The Grocer	Good coverage of food retailing and wholesaling trends.	Wm Reed Ltd 517 Southwark St LONDON SE1 1RQ *Tel:* 071 407 6981 *Fax:* 071 378 6781 Weekly
Fish Farming International	Limited coverage of marketing but often good articles about marketing and market trends.	AGB Heighway Ltd Alexander House 81–89 Farringdon Road LONDON EC1M 3LL *Tel:* 071 405 0500 *Fax:* 071 405 8555 Monthly
Fish Farmer	As for *Fish Farming International.*	34 Amberley Drive Woodham WEYBRIDGE Surrey KT15 3SL *Tel:* 0932 851668 *Fax:* 0932 859747 Bimonthly
Seafood News	Marketing journal of all types of seafood covering UK.	AGB Heighway Ltd Alexander House 81–89 Farringdon Road LONDON EC1M 3LL *Tel:* 071 405 0500 *Fax:* 071 405 8555 Monthly

USEFUL ADDRESSES

Sea Fish Industry Authority
Sea Fisheries House
10 Young Street
EDINBURGH
EH2 4JQ

The Scottish Business Information Service
Scottish Science Library
National Library of Scotland
33 Salisbury Place
EDINBURGH
EH9 1SL

See also Useful Addresses section at the end of the book for further information.

6
Distribution

Adding value in the chain
Choices
Opportunities and alternatives
The requirements of trade customers
Choosing trading partners
Trading locally
Running shops
Useful directories

ADDING VALUE IN THE CHAIN

Very few fish farmers deal directly with the general public. Most deal through trade customers, wholesalers, retailers or caterers. These people perform valuable services for both the fish farmer and the final consumer. They assemble supplies from many sources and then break down quantities into the smaller amounts wanted by individual customers. They enable fish farmers to sell their products in markets that would otherwise be inaccessible. The value of their services is represented by the margins which they take.

CHOICES

This chapter is about selecting trade customers. These are important choices because, unlike some other marketing decisions, they can be difficult to change if they are wrong first time:

- An important element in successful relationships with trade customers is building mutual trust and co-operation – this takes time.
- It is not always easy to switch partners if there are difficulties in the relationship.

The fish farmer has to choose:

- Types of trade customer.
- How many trade customers.

OPPORTUNITIES AND ALTERNATIVES

1 *Farm gate and local sales*
Sales to passing customers, local shops and hotels.

2 *Wholesalers*
There are a number of different types:

- *Trading companies*
  Buy and resell fish to other wholesalers or to retailers and caterers. Some in Scotland, some in England. Good companies have developed strong brand identities and operate in both home and export markets.
- *Inland markets*
  Found in major conurbations such as London (Billingsgate) and Birmingham. Made up of a large number of wholesalers whose main sales are to local shops and hotels but some sales to other wholesalers and processors.
- *Regional wholesalers*
  Buy relatively small quantities, supply local shops and hotels. Often combine wholesaling and retailing.

3 *Fresh fish to multiple retailers*
Supplied directly into shops by fish farmers or via specialist distribution companies. Normally no wholesalers used.

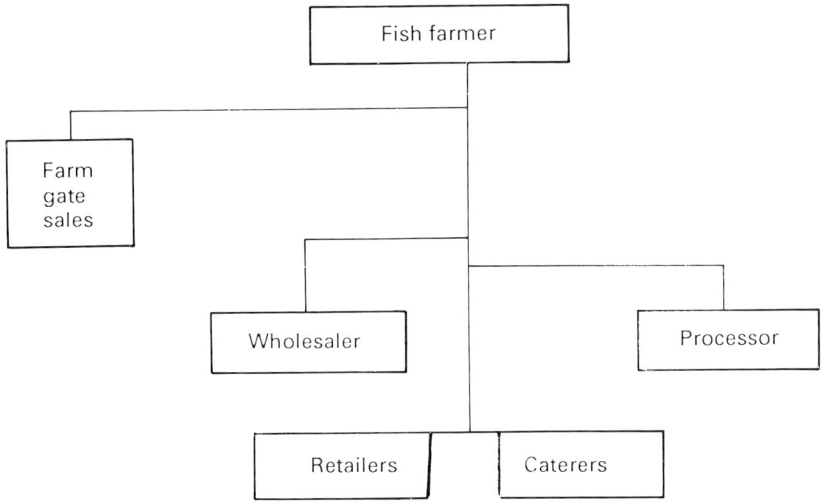

Distribution options

Fig. 6.1

4 *Packaged fish to multiple retailers*
Normally delivered to the chill depots of the retailers who carry out their own deliveries into shops.

5 *Processors*
Processors in home and export markets usually buy direct from farmers.

6 *Caterers*
Most sales of fresh fish are via wholesalers because caterers like to buy assortments of fish in small quantities and frequently. Wholesalers offer the best way of doing this. However some larger outlets do buy direct from farmers.

THE REQUIREMENTS OF TRADE CUSTOMERS

The structure of distribution channels varies between markets reflecting:

- Patterns of consumer demand.
- Locations of customers.
- The structure of fish farming and processing.
- The economics of distribution.

The suitability and requirements of the main trade customers are summarized below:

Customer	*Suitability*	*Requirements*
Farm gate	Only when good passing trade. Shopkeeping skills needed High margins on relatively small volumes – most suited for well-located small farmer	May be very seasonal depending on location. Range of related products an advantage.
Mail order	Mail order sales of fresh product little developed UK. Potential for sales of products which can go for periods without refrigeration.	Development of good mailing list. Potential for groups of businesses to develop joint sales.
Trading companies	Variable: closeness of relationships with farmers and length of commitment depends on the company. Outlet for large and small volumes.	Good companies concerned to offer customers high quality, consistent supplies and will want to work closely with grower. Some contracts but few with fixed prices.
Inland markets	Outlets for large or small volumes. Most wholesalers have regular suppliers but depending on conditions will take from fringe of other suppliers.	Most wholesalers prefer steady deliveries but some will take spot quantities at lower prices. Vary in quality requirements and price sensitivity but high consistent quality generally preferred. No contracts or fixed prices.

Table (cont.)

Customer	*Suitability*	*Requirements*
Regional wholesalers	Regional sales to fishmongers and caterers. Tend to have one or small number of regular suppliers.	Small orders. Weekly ordering. No contracts or fixed prices. Quality, etc. – as for inland markets.
Multiple retailers	Large volume orders. Likely to deal only with largest farmers or groups. Seven major companies. Most sales of fresh and packaged products. Prices move with market prices but less likely to fluctuate on a week by week basis.	Regular supplies on on-going basis. Require consistency and flexibility. High and carefully monitored quality standards. Informal agreements not contracts.
Other large retailers	Buy smaller quantities but direct links with growers especially in London area and in Scotland.	Regular supplies. Prices move closely with market prices. Ordering mainly week by week.
Smokers	Wide variety of size of business. Wide variety of quality standards. Vary in supply conditions: some want regular committed deliveries.	Large smokers have the same buying conditions as retail multiples except that they can hold stocks in frozen form if necessary (practices vary). Often very price sensitive. Some use of fixed price contracts.
Other processors	Wide variety of sizes.	Some have same specifications as multiple retailers – continuous ordering large quantities, high hygiene standards and high quality sought. Informal, non-contractual relationships.

CHOOSING TRADING PARTNERS

This list should help in the choice of trading partners, but it is only the beginning. Choices which you make must be looked at in great detail. Some general guidelines are:

1 *Look carefully at what each option offers*
Each trade customer offers something. Lower prices to some outlets can be compensated by lower costs of supplying them. For instance, selling to processors or a small number of local wholesalers may save you extra selling and administrative costs which compensate for lower prices. Higher prices can involve higher costs, e.g. costs of delivering to more distant destinations. Check this out. Check out the true costs of supplying different types of accounts including:
- All your time,
- All the transport,
- All the administration costs.

2 *Be committed*
Make sure that you are fully committed to the options you choose. For instance, you may wish to build up accounts with small regional wholesalers but if you do this, there are likely to be advantages in building up a number of these types of account so that you can set up an efficient system for serving them. Don't try to spread sales across too many customers all with different requirements and no common cost economies.

3 *Spread*
Unless you have very close and secure relationships, some limited spread of customers is a good idea because it means that you will have fewer problems if you lose any one of them.

4 *Invest in relationships*
Seafood traders are more and more likely to want close relationships with suppliers in order to guarantee their own services to customers. This means spending time developing this and not flitting from account to account.

5 *Avoid outlets where you cannot compete*
Everyone talks about the multiple retailers – but they demand standards of consistency, flexibility and order sizes that small fish or shellfish farmers cannot offer. Leave them to bigger competitors or join a marketing group yourself.

6 *Think about groups*
For most of these customers, marketing groups or larger businesses have advantages. Think again carefully about the advantages of becoming part of them. If this is not possible, think about other partners. If you want to supply Scottish local wholesalers, for instance, is there someone with other products who could market them with you and save transport and selling costs?

TRADING LOCALLY

Much farmed fish, particularly shellfish, goes to Scottish customers. This is very helpful because it provides a firm local base on which the business can be built.

Many of these accounts can be supplied just as well or even better by small businesses:

- Personal contact
- Flexibility
- Easy contact

But there are other local fish farmers who may want the same business so make sure you offer the same high standards of customer care that you offer other customers. Don't take these local customers for granted.

RUNNING SHOPS

If there is a good passing trade, an option to be considered is operating a farm shop. Before starting check:

- *What will attract people to the shop?*
  Is it enough to sell one product or do you need a range of food products?
- *Which is the best location?*
  At the farm is the easiest and this is attractive for customers because it is interesting. If you move elsewhere costs will be higher but this could be compensated by extra business. Check numbers of passing traffic.
- *Costings*
  A farm shop must be profitable in its own right. Make sure that your time is properly costed. Will the shop enhance the total business or will it adversely affect the management of the farm itself? How much extra has to be sold and at what prices to make this a better option than selling to wholesalers?

WAYS TO SUCCEED WITH FARM SHOPS

DO

- Make sure it is spotlessly clean *always.*
- Make sure product is attractively displayed – this does not mean very expensive shop fittings but does mean that a bright and attractive environment is created.
- Keep the shop tidy – *always.*
- Make sure that the products are kept at the right temperatures and *never* sell when they are poor quality.
- Make sure that staff are attentive, pleasant and know something about the products.

- Make sure that customers are directed to the shop by attractive, professional signs.
- Make sure that customers know about the shop through tourist associations and other publicity.
- Make sure you have all the necessary planning permissions.

DON'T
- Run the shop as a convenience to yourself, not customers.
- Give an untidy, messy impression with poor products. One of the reasons why so many small shops have gone out of business is because they were not very good shops.
- Compete with the local grocer – you can't. Select other products similar to your own which convey the same kind of impression and which make you different.

ACTION POINTS

Have the costs and benefits of all customer accounts been checked over the past year?

....

What is the plan for customer accounts for the next year?

....

If this involves changes:

1. List what the business has to do to make them successful.

   ....

2. Estimate the revenues and costs compared with alternatives.

   ....

3. List actions being taken over the next year to improve the service offered to customers.

   ....

USEFUL DIRECTORIES

Fish Trader Handbook (annual)
FMJ International Publications
Retail Journals Division
Queensway House
2 Queensway
REDHILL
Surrey RH1 1QS

European Food Trades Directory
Newman Books Limited
48 Poland Street
LONDON W1V 4PP

Frozen and Chilled Foods Yearbook
Marketing Department
Room 302
FMJ International Publications
Queensway House
2 Queensway
REDHILL
Surrey RH1 1QS

Seafood International Directory (annual)
Alexander House
81–89 Farringdon Road
LONDON EC1M 3LL

7
Doing Business with Customers

INTRODUCTION

Good products deserve effective promotion. Products do not sell themselves but need carefully thought out plans for talking to customers and persuading them to buy.

THE ELEMENTS

There are a number of different ways of talking to customers:

- *Selling*
  Talking to customers in person.
- *Advertising*
  Paid space in trade journals, magazines, TV, exhibitions, leaflets, etc.
- *Publicity*
  Favourable news coverage.

In general in a small business the emphasis is on the first of these, *selling and the personal approach*, because:

- You can't compete on equal terms with the big businesses on advertising so it is not cost effective to try.
- A personal approach which involves direct contact and a two-way dialogue with customers can be just as effective.
- Probably only your standards of service distinguish you from your competitors. The sales contact is the listening post for problems and ways in which levels of service can be improved.

The exceptions are some forms of advertising which do work for small businesses such as business directory listings, and small ads in trade magazines, promotional leaflets and exhibitions at trade shows. These are discussed later.

There is one more exception. The small business alone does not have the resources to reach final consumers. Small fish farmers together can however do this through the trade association and this will help to 'pull the product' through from the consumer end. Read Chapter 10 to see why this is so important and necessary.

APPROACHING NEW CUSTOMERS

Approaches to new customers can be made in a number of ways:

- By letter and follow-up phone calls;
- By letter and follow-up visits;
- At exhibitions;
- Replies to enquiries.

Sending sales literature in advance prepares the way for discussions and indicates to the buyer that the approach is a serious one. It is sensible to follow this up by phone to establish whether the prospect is worth pursuing.

Decisions have to be made as to the balance of personal visits and telephone contacts. Each has its uses:

PERSONAL VERSUS TELESALES

	Personal selling	*Telesales*
Uses	Selling. Order taking. Maintaining goodwill. Notifying customers of major changes. Following up major service problems. Maintaining contact.	Looking for new prospects. Making appointments. Inbound marketing (e.g. freephone). Selling. Taking orders. Maintaining goodwill. Notifying customers of changes. Providing customer service. Maintaining contact.

Advantages	Detailed presentations. Easier to establish relationships Better two-way communications flow.	Saves time. Saves money. Convenient. Flexible.
Shortcomings	Costs. Time – yours – your customer's.	Cannot appeal to senses of touch, sight or taste of product. Easier to say no.

Much fish business is managed entirely by telephone. At the same time, for major accounts it is worth considering how some personal contact can be made because:

- Face to face contact is more persuasive and gives more information to both buyer and seller.
- It establishes a closer personal relationship which will make things work more smoothly.

Face to face contact can be very expensive, especially for overseas markets but it should be regarded as an investment in building a business just like any other. There are a number of ways in which costs can be held down:

- Go to exhibitions at which buyers are present.
- Plan schedules so that a number of customers are visited each trip.
- Get them to come to see you.

SELLING: EFFECTIVE PRESENTATIONS

It is worthwhile spending a lot of time on the presentation to be made to customers:

1 *Do research*

Find out as much as you can about prospects from:
friends,
business contacts,
trade directories,
exhibitions and catalogues,
ads in journals.

Some prospects will be more suitable than others – don't waste time on those that are not.

2 *Find out about the business*

Any information you can find out in advance will be useful. Try to find out about turnover, their main interests, reputation, quality standards, customers. Check out people's credit worthiness before you spend a lot of time building a contract.

3 *Do some homework on their markets*
Find out what has been happening to the market, how the products are sold and where they are sold. This is fairly easy for food products because you can have a good look at shops and restaurants.

4 *Present benefits clearly*
Plan carefully in advance – whether it is by phone or by personal interview.

Use the sales contact form shown in Fig. 7.1 for planning.

Name of firm ____________________

Person with buying authority ____________________ Title ____________

Date of visit ____________________

1. What do I want to accomplish from this call?

____

____

2. Previous call/contacts
Dates and results ____________________

____

____

3. Competitors selling to this firm ____________________

____

4. Key points to make in presentation ____________________

____

____

5. Objections anticipated and counter arguments

____

____

6. Method of asking for order ____________________

____

____

Fig. 7.1. Planning sales contacts.

MAKING EFFECTIVE PRESENTATIONS

1. Relate offering to customer needs.
2. Provide reassurance on quality, delivery, etc.
3. Show complete product knowledge.
4. Show complete knowledge of logistics.
5. Demonstrate thorough knowledge of financial issues relating to sale.
6. Offer opportunities for discussion.
7. Be realistic in your goals.
8. Don't make promises you cannot fulfil.
9. Take notes about what was discussed.
10. Evaluate the results of each call and decide the next action.

PRESENTING YOURSELF

1. Dress in a business-like manner.
2. Always identify yourself, your company, your purpose.
3. Speak clearly and at an understandable rate.
4. Be courteous, never argumentative or insulting.
5. Show enthusiasm and confidence.

EFFECTIVE NEGOTIATION

Because of the similarity of products, it is often said that the fish farmer has little discretion in negotiation. But price surveys show that there are variations between farmers. There are a number of explanations, but undoubtedly one is the skill shown by different growers in negotiation. This should not be overstated because a small seller undoubtedly has relatively little market power compared with a large buyer. Nevertheless good negotiators are often able to get some improvements on the initial offer.

Naturally the buyer will try to get a better deal than the seller wants to accept. In the process of negotiation each will make concessions in an attempt to bridge the gap. Each will have terms which they cannot accept, but if a deal is to be possible, there will be a 'grey' area in between. Effective negotiators try to secure as much of this area for themselves as possible. They are more likely to succeed if they follow some simple rules:

THE ART OF NEGOTIATION

1 Aim higher and you come out higher.
2 Use knowledge – the more you know about your business, your competitors and what is going on in the industry, the better you will be in negotiation. Spend time finding out what deals other people are making, participate in price surveys.
3 Remember that competitors are not necessarily offering a better deal, whatever the buyer says.
4 Be sure of the minimum terms acceptable to you before you start and stick to them.
5 Spot tactics as tactics – don't believe everything they tell you.
6 Be committed and believe in your business and what it can offer.
7 Be sure you know the cost of all services – be careful not to make extra offers that you have not costed.
8 Remember that you are trying to build a long term relationship – the customer you have had for years deserves as much attention as the new customer.

SELLING: MAINTAINING THE RELATIONSHIP

Many contacts with customers are not pure selling ones. They include routine administration of orders and technical discussions. Within an established relationship, these may be the main contacts and it is to your advantage to look after them carefully.

The rules are simple and mirror guidelines elsewhere in the book. What is necessary but more difficult is to ensure that they are followed by everybody:

- Be a service fanatic.
- Make sure that feedback from customers is communicated to the rest of your organisation.
- Respond to criticisms, don't become complacent with praise.
- Try to find out how the customer expects their trade to develop so that you have a better understanding of their needs.
- Don't waste the customer's time – contact only as necessary but do try to meet occasionally.

- Handle any problems or misunderstandings as courteously as possible, even if you think you are in the right!
- Watch for signs of deterioration in the relationship, e.g. reduced ordering, more complaints. Establish the cause and if possible solve it before position deteriorates to the point where customer is 'not available' on the phone.

ADVERTISING: THE BASICS

As a minimum all businesses should have:

- *A leaflet describing the business and its products.*
- *Headed notepaper.*
- *Business cards.*
- *Packaging.*

These should be related to each other by the same design themes. Producing to good quality standards is not cheap but is a worthwhile investment. It is recommended that the design should be done either by an advertising/design agency or by a printer with a reputation for good design work. A well-designed brochure reinforces the image of the business which you want to give the customer. You may have good ideas but get a professional to implement them for you. Don't do it yourselves!

TRADE ADVERTISING

You may want to consider:

- Listings in business directories.
- Small advertisements in trade magazines.

Whether this is worthwhile will depend very much on your objectives and, in particular who you are trying to reach:

- *Targets*
  Are you trying to reach small numbers of customers who could more easily be reached by letter or phone or visit, or are you trying to reach a lot of people so that advertising is the more cost effective route?
- *Message*
  Do you need to tell people you exist? Do you know who you are trying to tell? If you want to tell new people about yourselves but you don't know them, advertising is a good way of getting them to come to you.

In general, it is worthwhile making sure that your name is in the relevant business listings. Often this is not very expensive and it is easy to do but make sure that it is:

- *The right directory*
  Check out other entries to see that it is reaching the kind of customers you want. Ask the publishers for details of their market.

- *Up to date*
  It creates a bad impression if what people read turns out to be not quite correct when they contact you.

You may also want to consider whether to take a larger display ad within the directory or whether to take space in a trade magazine. Again there are some useful guidelines:

- *Check out readership*
  There is no point in talking to people who are not interested – the magazine will give you details.
- *Check out costs in relation to coverage*
  What matters is not just the total cost, but how effective it is. Check out costs, not in relation to total circulation of the magazine but in relation to circulation among the type of people you want. The magazine will help.
- *Start small*
  Start at a price level that you can afford. It won't put a strain on your finances. If you do get a positive response, you can always increase the size or amount of advertising next time round.
- *Messages*
  Think carefully about what you want to say (see below).

GOOD ADVERTISING

Remember that people forget easily and ignore as much as they absorb. Advertisements are intended to inform, interest, stimulate or prompt action – not as a catalogue of every feature of the business and its products. Some helpful rules to follow are:

1 Stick to one key idea – don't try to say too much.
2 Try to be attention grabbing in the designs and words chosen.
3 The business name or that of its products must register strongly.
4 There must be some benefits for the customer.
5 Avoid jargon or too many technical words.
6 Try to reflect the personality of the business.

FOLLOWING-UP

The advertisement or listing is not the end of action. If they work, people will want to contact you and you must be ready for them:

- *Back-up telephone answering machine*
  This is essential. Calling a business and having no one answer puts people off. Follow these rules:
    Let people know approximately when you will be returning their call.
    Buy a machine that lets the caller talk for as long as they need to.
    Keep messages on the machine up-to-date.
    Get a good machine and fix it immediately if it breaks.
    Get one that you can review by telephoning your own number.
- *Fax*
  More and more businesses use fax because they find it easier than posting letters. See if you can afford one.
- *Respond!*
  Make sure that you do respond to enquiries as soon as possible and ask enquirers how they heard about you. This lets you know which forms of promotion are proving effective.

It is also helpful to make sure that you keep a record of contacts so that you can make the most of those contacts that you already have (see Fig. 7.2).

EXHIBITIONS

Exhibitions are another useful way to:

- Make new contacts.
- Strengthen existing contacts.

Exhibitions can be expensive but there are often opportunities to join other farmers through your trade association or through organisations such as HIDB/HIE. But exhibitions are hard work and it is important that you get the most out of them:

- *Display*
  Your display should be eye-catching and attractive. It is better to have a small but effective display than a large but poor one. This is where you need to spend on display material – a proper display stand is not very expensive, especially if you can re-use it. It should always be clean and tidy – just like the shop (see Chapter 6).
- *Be active*
  Take a lively interest in passing visitors and get them to take an interest in the stand. Discourage your friends from visiting you when you should be looking for customers and don't eat or smoke on the stand. If at all possible try to make sure that there are at least two of you so that one can be away from the display making other contacts and encouraging visits.

	6 mo. ago	1 yr. ago	2 yrs. ago	3 or more yrs. ago
Your marketing list was last updated	☐	☐	☐	☐

Is a complete list of your customers, suppliers, prospects available for an immediate mailing?

☐ Yes

☐ No

		excellent	adequate	needs more work
If 'Yes', how current are your addresses:	customers	☐	☐	☐
	suppliers	☐	☐	☐
	prospects	☐	☐	☐
How current are your phone numbers:	customers	☐	☐	☐
	suppliers	☐	☐	☐
	prospects	☐	☐	☐

If 'No', how long would it take you to make a list of:

customers ________ hrs

suppliers ________ hrs

prospects ________ hrs

Starting now you are compiling these records from:

☐ payments

☐ customer records

☐ organisation membership lists

☐ your personal address book

☐ other ____________________

Fig. 7.2. Your marketing list.

- *Be organised*
  Make lists of people you want to see at the exhibition and be systematic about trying to see them.
  Fill in sales contact reports (Chapter 5). Follow up interest after the exhibition as soon as possible.

There are several good books available outlining how to get the most out of trade exhibitions. Ask HIDB/HIE.

PUBLICITY

This means getting favourable news coverage in trade newspapers and magazines. It is unpredictable:

- It can produce an influx of one-time customers. If you can't handle them, long term customers may be alienated and the business harmed.
- Or it can produce very little return for the effort.

But if it is used intelligently it can get your business known to new customers and reinforce the trust of your existing customers:

- *Have a story*
  Unless you have some interesting news, forget it. Magazines will not be interested if all you want is self-promotion with nothing of interest to their readers.
- *Be honest*
  There are big dangers in telling anything other than the strict truth – you are likely to be found out.

If you have something interesting to say, it is worthwhile contacting magazines likely to be read by your customers. If the journalists are interested they will follow-up the story and tell you what they want, though you may have to take some advertising space to obtain editorial.

ADVERTISING AGENCIES

Considerable help can be obtained from a good agency who, as well as designing materials, should be a source of advice. There are successful companies in the fish business who carry out all their own promotional work but these companies are run by highly trained marketing professionals.

If you do decide to use an agency, choose carefully. Agencies vary in experience and the type of companies whose accounts they handle. Some guidelines are to select an agency which:

- Has experience in advertising food products.
- Has experience working with small and medium size businesses.

- Is willing to work realistically with businesses who have only limited money available.
- Appears to have a genuine interest in working for you.
- To whom you can talk easily.
- Who are not carrying accounts of your competitors.
- With a proven track record of success.

The right agency in turn has a right to expect a lot of things from you. A successful relationship is a two-way one. They must know about your business and they must know about your marketing objectives. The more information you give them, the more effective will be their work. They will need very clear objectives so that they know exactly what it is they must do for you.

SUMMARY

Good communications are vital for successful selling. This chapter has shown that they are not the monopoly of the large firm and that there are many cost effective ways available to smaller businesses.

8
Exporting

INTRODUCTION

The principles of successful exporting are the same as for marketing in the United Kingdom. The difference lies in complexity: markets are less familiar, there are different preferences, languages, customs and regulations. This chapter looks at some key issues for successful exporting and offers advice on documentation and financing.

WHY EXPORT?

Exporting offers a number of advantages:

- An expansion opportunity.
- An opportunity to increase the number of customers and markets, and to spread risks.

- Possible opportunities for greater profit (although this is not always the case!).

But there are difficulties:

- Export markets are more unstable because of fluctuations in exchange rates and because (especially for major markets) they buy fish from many different sources, with possibly a weaker preference for the Scottish products than in the UK.
- The time and money costs of marketing to export markets are likely to be higher than to the home market.

Advantages can offset the difficulties, particularly for higher value fish products. This has been shown by the rising exports of Scottish farmed fish over the past ten years but the extra difficulties and risks involved mean that the decision to export should *not* be lightly taken.

SHOULD YOU EXPORT?

Many Scottish companies export fresh and processed fish. They vary from multinational companies to small family businesses. A number have been so successful that exports dominate their sales. This success has been based on:

- *Enthusiasm, motivation*
  It takes time to build contacts and establish satisfactory relationships. Results cannot be expected quickly.
- *Commitment and resources*
  It is easy to export when supplies are short but as competition intensifies, much more attention must be paid to customers:
  The need to use their language which is often not English.
  Willingness to visit customers and be at trade exhibitions.
  Willingness to manage more complex trading and documentation.
  Willingness to trade in different ways to different types of customers.
  The need for different packaging and labelling.
- This may involve employing new specialist staff. A number of Scottish companies have already taken this step and have multilingual staff who work easily in overseas markets:
  'I sell in English but I buy in German' – German manager
  - Can you supply the sizes, qualities and volumes required?
  - What advantages do you have which will make you more effective than competitors?
  - Are you prepared to invest the time, money and commitment to succeed?
  - What gaps exist in your capabilities and can you overcome them?

In general for fresh fish and shellfish, larger companies are more successful than small companies in export markets. They can supply regularly and in volumes which give transport and handling cost savings. For smaller companies, rather than attempt to export directly it may make more sense to join a marketing group or to link with good Scottish traders who are more experienced and handle larger quantities.

With processed products successful smaller companies have specialist high quality products fitting particular market niches. To be successful in competition with larger companies, small companies should be aware of the need for distinct non-price advantages over other products. Without them exporting is likely to be difficult and unprofitable.

PASSIVE AND ACTIVE EXPORTING

Many companies export from time to time in response to enquiries and this is known as *passive* exporting. This is fine but don't confuse it with *active* exporting where a company makes a positive move to create a long-term export business. Don't let the number of export orders build up until they are taking up a lot of time and effort but in a haphazard and non-cost effective way. You will end up with a spread of customers across markets and high costs.

Proper effective exporting means making plans for:

- Markets to be chosen.
- Trade customers to be selected.
- Marketing actions.

SELECTING EXPORT MARKETS

Export markets are not all the same, even within Europe. For the new exporter, drawing up a list of possible markets is a process of balancing:

- The size of the opportunity.
- Levels of risk.
- Levels of competition.
- Distribution costs.
- Match with company capabilities.

A checklist of possible questions is shown below.

SELECTING EXPORT MARKETS

The opportunity	Size of market.
	Expected growth.
	Ease of access.
	Suitable trade customers.

Selecting Export Markets (cont.)

The risk	Currency fluctuations. Currency convertibility. Level of inflation. Government intervention. Political instability.
Level of competition	Price levels. Number of competitors. Cost advantages/disadvantages of competitors. Market preferences for different sources of supply.
Company capabilities	Level of sales sought. Capability of meeting product and service requirements. Levels of marketing support needed. Modifications to normal trading practices needed.

All these points must be considered, but there is no simple right answer – it is a question of balancing issues. For instance, the French seafood market is large, still growing, easily accessible and a profitable export market for a number of companies. Markets in the Middle East are more difficult to access, smaller and more risky, but they are less competitive and prices may be higher.

Which market you select will depend on:

- The capabilities of the company.
- The balance of risk and opportunity sought.
- Many companies develop a number of markets to provide a balance of risk and opportunity but it is suggested that you start with just one market.

Again, having large scale operations is an advantage.

RESEARCHING MARKETS

There are a number of sources of valuable information:

- Government departments.
- Trade associations.
- Chambers of commerce.
- Banks.

- Reports and studies by market research firms, universities and research institutes.

Lists of relevant publications can be obtained from the Scottish Export Office and from the Export Intelligence Library of the Department of Trade and Industry in London.

The export services of the Department of Trade and Industry include the following:

- Finding agents and business contacts.
- Providing information on import regulations.
- Export market research.
- Arranging overseas visits, trade visits, store promotions, and trade missions.

SELECTING TRADE CUSTOMERS

The criteria for selecting customers are the same as for the United Kingdom. There will be different types of customer and the exporter has also to select the individual companies with whom they want to deal.

For fresh fish the most usual method has been to sell to importer/wholesalers or processors. Direct sales to regional wholesalers, retailers and caterers are now also growing rapidly.

AGENTS

Agents can be useful, particularly for value added processed products, but commission agents are also sometimes used for fresh products. The commission agent represents the exporter in the overseas market and seeks orders on their behalf. This is a useful service when there are many small accounts and when the exporter is not familiar with the country.

Sometimes there are problems. Trade customers require a fast responsive service and an extra link in the chain can slow things down. With modern communications a link direct to these customers may work more smoothly. Assess how important it is to have continuous presence in the market and the quality of the proposed agent when taking this decision.

Choose agents carefully because they become your company in the eyes of customers. Look for agents who are:

- Selling to the right kinds of outlets for your type of business.
- Selling other products which are compatible and not competing.
- Technically competent and able to provide the right kinds of services.
- Of good financial standing.
- Enthusiastic and willing to learn about you and your products.

Normally the initial agreement is for a trial period. When setting up the agreement, make sure that you give the agent objectives and that you both agree on these and on marketing methods and effort.

EXPORT LOGISTICS

Export is a group activity involving:

- *Fish farmer/processor*
  Packing of product.
  Some documentation.
  Selection of carrier.
  Selection of freight forwarder.
- *Transport companies*
  Movement of goods to destination or to freight forwarder.
- *Freight forwarder*
  Services include arrangements for international collection and deliveries, handling and provision of documentation, insurance, groupage services.
- *Customers or their representatives*
  Import clearance, arrangement for collection.

For exports by land and sea, the transport company moves product either to shipment points or directly to customers, with freight forwarders responsible for documentation at customs and shipment points. Exports by air normally travel by road to the air freight terminal as the responsibility of the transport company. They are handed over at the air freight depot to the freight forwarder who handles documentation, customs clearance and supervises loading of cargo by the airline.

Use freight forwarders with a proven track record in handling perishable cargoes. Transport companies must meet tight time deadlines with valuable cargoes and should also be able to demonstrate a track record of performance.

EXPORT DOCUMENTATION

Export documents are necessary for:

- The movement of product.
- Invoicing the customer.
- Receiving payment for the products delivered to the customer.
- Import and export government regulations.

Documentation has been simplified in recent years by computerised documentation methods, particularly for exports to the other members of the European Community. The freight forwarder can take much of the difficulty and effort from the fish farmer. But inaccurate documents cause expensive delays to perishable cargo and it is extremely important that all relevant information is provided in the correct form.

The following list of required documentation is only an indication. Regulations change and should always be checked. A good freight forwarding company and the Department of Trade and Industry can provide invaluable advice.

1 COMMERCIAL INVOICE

Contents	Name of exporter, Name of consignee, Invoice number and date, Country of destination, Description of cargo, Value of cargo, Payment terms, Name of freight forwarder, Signature, Conditions of delivery.
Organisation	Should accompany the consignment and multiple copies required by banks and government authorities as well as the buyer. Normally a copy of the invoice will be sent under separate cover by the exporter to the customer.

The normal currency of invoicing for fish exports is Sterling. This is probably best from the point of view of the seller since it avoids the risk of loss due to currency fluctuations. However, the currency used should be dependent on the wishes of the buyer. Banks can give advice on ways of handling the risk associated with invoicing in other currencies.

There are a variety of types of contract. Normally invoices used by air are 'CIF' (cost, insurance and freight) invoices. The price is the cost of goods loaded plus cost of freight and insurance to the agreed point of delivery in the buyer's country. Customs clearance is normally paid by the buyer, as are delivery, taxes and final distribution costs. Where goods are exported by road on roll-on roll-off ferries, they are more likely to be delivered 'CIP' (carriage, insurance and freight paid to named destination) with all costs of delivery paid to the buyer's address but excluding duty payable which is the responsibility of the buyer. Other variations are possible so that it is necessary to reach precise agreement on the charges to be paid by the exporter at the time when the deal is negotiated.

There are standard documentation forms available, which should be used. The forwarder can advise.

2 CERTIFICATES

Certificates of health

Health certificates are the requirements of the importing country and not the United Kingdom. Thus the health certification requirements can and do vary from country to country. They change as disease incidence changes and as attitudes change towards particular issues. In the European Community regulations are currently undergoing revision as part of the creation of the European Single Market.

It is the responsibility of the exporter to understand and meet the requirements for the country of destination. Information can be obtained from the relevant consulates and embassies. *Bone fide* exporters can also seek advice from:

- Department of Trade and Industry (Export Desks).
- Department of Agriculture and Fisheries for Scotland (DAFS).
- Local Environmental Health Offices.
- Trade Associations.

Certification is not necessarily available immediately on demand. For instance, to demonstrate freedom from disease evidence may be required over 18 months or longer that a farm has been disease free. Certification may relate to the conditions in packing and processing plants as well as the product itself. Further, with increasing international concern over food standards, health requirements are likely to become more stringent in the future.

Certificates are normally issued by Local Environmental Health Officers with the exception of certificates relating to the absence of pathogens where the advice of DAFS should be sought. Certificates should always be provided by official bodies. Testing by private agencies does not carry the same status.

Failure to provide necessary health certification can have serious consequences since consignments may be rejected on arrival or remain in customs for long periods leading to severe deterioration of contents.

Certificates of origin

Some importing countries require certificates of origin of the goods. These are provided by the local Chamber of Commerce who certify the origin by endorsing the invoices, usually with a stamp.

Certificate of insurance

It is necessary to insure consignments in transit. This is the responsibility of the exporters and the insurance certificate normally remains in their possession, although occasionally buyers will wish to see a copy.

3 CUSTOMS DOCUMENTS

European Community 'T' Forms

This standard document is used to control the movement of goods at all stages within the EC and has replaced the need for separate customs forms for each country. It shows that duty has already been paid and that the goods are able to move freely within the European Community. This document will be prepared by the freight forwarder or the shipping agent.

Customs export declaration

This is prepared by the freight forwarder or shipping agents. It is a single administrative document for the movement of goods through customs.

4 TRANSPORT DOCUMENTS

Air waybill

The freight forwarder prepares the air waybill. This is a contract between the forwarder and the airline to move the consignment. It is non-negotiable and not a document of title. Multiple copies travel with the consignment.

Bill of Lading

Maritime bills of lading have a similar role to air waybills but for goods which are moved by sea. They are prepared by the shipping agent on behalf of the exporter. There is a difference between a maritime bill of lading and the air waybill since the maritime bill of lading is a document of title which means that the company named on the bill of lading has the right to possess the goods. This facilitates arrangements for payment for frozen goods which are in transit for long periods.

4 Transport Documents (cont.)

Transport Bill of Lading

This documentation accompanies goods exported by roll-on roll-off systems. It contains similar information to the other transport documents and is prepared by the freight forwarder.

PAYMENT AND FINANCE

There are many methods of financing exports, including government support systems but most payment systems for fish are relatively simple. Business is normally based on regular trading and straightforward systems are needed.

Most trading is 'Open Account' which means that the seller trusts the buyer to pay on the terms agreed, either on receipt of the consignment or within a negotiated number of days. Normal bank loan finance is sought if needed by the exporter to cover the delay before payment is received.

The main difficulties with this system are in the early days of dealing with unknown buyers. On the whole the major international traders with whom exporters deal behave responsibly and agreements are honoured, although there have been some examples of failure to pay. The commercial departments in British Embassies can help with research on the financial status of customers, and credit ratings can be obtained from Dun and Bradstreet for a fee. Initially it is prudent to start trading with small consignments.

Letters of credit are occasionally used to reduce risk with unknown buyers. The bank of the importer undertakes to pay an exporter for the goods providing that they comply with the conditions laid down in the letter. This ensures that the exporter receives payment for the goods and that the importer does not have to pay until stipulated conditions have been fulfilled.

EXPORT CHECKLIST

List the reasons why you think that you can export successfully

1 ....

2 ....

3 ....

4 ....

5 ....

Have you checked for each selected market?

	Market (1)	Market (2)	Market (3)
Size	....	....	....
Ease of access	....	....	....
Growth potential	....	....	....
Ease of distribution	....	....	....
Risks	....	....	....
Levels of competition	....	....	....

Describe your marketing strategy:

1 Objectives ....

2 Customers ....

3 Product ranges ....

4 Promotion plans ....

5 Distribution ....

(Repeat for all markets)

Have you checked that you have?

1 Correct invoicing procedures ....

2 Necessary health certificates ....

3 Other certificates (if needed) ....

4 A reputable freight forwarder ....

5 Reliable transporters ....

6 The most cost effective transport ....

7 Creditworthy customers ....

(Repeat for all markets)

USEFUL ADDRESSES

Department of Trade and Industry
Victoria Street
LONDON SW1
Tel: 071 215 7877

Scottish Export Office
Industry Department for Scotland
Alhambra House
45 Waterloo Street
GLASGOW G2 6AT
Tel: 041 242 5493

Department of Agriculture and Fisheries for Scotland
Marine Laboratory
PO Box 101
Victoria Road
ABERDEEN AB9 8DB
Tel: 0224 876544

Dun and Bradstreet
Business Information Services
11th Floor
Centre City Tower
BIRMINGHAM B5 4UN
Tel: 021 631 2323

Euroinfocentre
Bothwell Street
GLASGOW G2 6NR
Tel: 041 221 0999

Royal Bank of Scotland PLC
International Division
PO Box 106
Moncrieff House
69 West Nile Street
GLASGOW G1 2LT
Tel: 041 248 2777

Bank of Scotland
International Division
8 York Place
EDINBURGH EH1 3EP
Tel: 031 243 5089

NatWest Bank
International Office
14 Blythswood Square
GLASGOW G2 6AT
Tel: 041 221 1838

Clydesdale Bank PLC
International Services
150 Buchanan Street
GLASGOW G1 2HL
Tel: 041 248 7070

Institute of Freight Forwarders
Suffield House
9 Paradise Road
RICHMOND
Surrey TW9 1SA
Tel: 081 844 2266

British Food Export Council
301–344 Market Towers
New Covent Garden Market
LONDON SW8 5NQ
Tel: 071 622 0188

See also Useful Addresses section at the end of the book for further information.

USEFUL BOOKS FROM THE DTI
(available from Scottish Export Office)

Department of Trade and Industry, Hints to Businessmen
Department of Trade and Industry, Export Handbook
Department of Trade and Industry, How to Start Exporting

9
Handling the Product

INTRODUCTION

Poorly handled fish arriving at customers in second-rate condition can undo years of work on other aspects of marketing. High standards of control and of hygiene are therefore essential in growing, harvesting and processing fish. This chapter looks at the issues involved in being a quality producer and gives some advice on the selection of transport.

QUALITY STANDARDS

Your quality standards must be set to meet:

- The demands of customers.
- Legal requirements.
- Comparisons with standards of other foods.

Details of legal standards for the harvest and post-harvest handling of both fresh and processed farmed fish will in the future be defined by a common set of standards throughout the European Community. The Department of Agriculture and Fisheries for Scotland (DAFS), trade associations or HIDB/HIE can be asked for details.

But legal standards are often not enough. Customers may have their own set of standards which may well be higher than legal standards. Trade associations (see Chapter 10) have been very concerned that the products of members compare well with other foods and have drawn up their own higher standards. The concern of members of the public about the standards of the food they eat also ensures that everyone's standards are changing – and are becoming increasingly strict.

SETTING STANDARDS

The details of quality control schemes and codes of practice can be obtained from your trade association. You can find the addresses at the end of Chapter 10. The requirements and coverage of different schemes vary so

You are strongly advised to contact the relevant trade association for further details.

Because of these differences and because some elements of schemes are confidential to participants, the list below is just a checklist of issues that you ought to find out about. The trade associations will provide you with detailed instructions or advice as to where you can get further information.

FRESH FISH

Are your fish starved properly prior to harvesting?

Are you stopping medical treatment long enough before slaughter?

Are your fish the right
- shape?
- colour?
- texture?
- smell?

Are your fish undamaged?

Are your fish slaughtered correctly in line with legal requirements?

Are they brought quickly enough to the right temperature for packing?

Are they the right temperature when boxed?

Are they packed quickly enough after slaughter?

Are they packed carefully?

Do you use enough ice?

Is your water of the required quality standard?

FRESH SHELLFISH

Have the shellfish been cleaned and graded properly?

Are they undamaged?

Have the shellfish been trained (where appropriate)?

Have they been packed and labelled properly?

Are they transported at the right temperature?

PROCESSING FISH

Are fish cleaned properly?

Is the internal appearance of the fish acceptable?

Are they cut correctly and neatly?

Are they packed properly?

Are they uniform in appearance?

FREEZING

Are you using the correct freezing methods?

Are you using facilities of the right standards?

Are freezing temperatures correct?

Are fish frozen quickly enough?

Are they properly packed?

Are they properly labelled?

Are they held at the correct temperatures for thawing?

Do they have the right appearance when thawed?

SMOKING AND OTHER PROCESSING

Are your facilities of the right standards?

Do you *always* keep them to the right standards of cleanliness?

- Staff,
- Equipment,
- Preparation,
- Packing.

What care do you take to avoid contamination?

Do you properly control salt uptake and smoke uptake?

Is the product of good appearance and condition?

Are products properly stored and sold well within their product lives?

Are products properly labelled?

BOXING

Do the boxes meet all standards of cleanliness required?

Are they free of abrasive surfaces?

Are they as defined by your quality control scheme?

What systems of inspection do you use?

There are a number of different types of boxes available. Polystyrene is most frequently used and robust heavy duty boxes are much safer in all situations. In addition your trade association may specify the types of boxing necessary to meet their quality control requirements.

It is especially necessary to use heavy duty boxes when:

- Delivering to UK wholesale markets.
- Delivering small consignments.
- Using surface transport outside the UK.

Light boxes can be used for UK deliveries where you have great confidence that the consignment will be handled carefully and when distances are short. It is necessary to check that they are satisfactory in use – don't just hope that they are all right!

Melt water has to be coped with. The solution depends on destinations and types of transport:

- *Mixed loads UK*
  Boxes must be sealed and watertight to prevent the escape of water which would damage other cargo.
  They must reach their destination in 24 hours or less.
- *Specialist fish transport*
  Normally boxes will have drainage holes to allow melted water to drain into the truck.
- *Air freight*
  If boxes are in transit for a longer period of time the problem of melt water is more serious. A variety of solutions have been found but the most successful using conventional ice has boxes with indentations on the underside on to which an absorbent pad is fitted for the melt water. The whole box is covered with a cardboard outer. In some cases fish and ice are bagged separately in taped boxes. An alternative is the use of gel packs which are significantly lighter than conventional ice.

TEMPERATURE CONTROL IN TRANSIT

Proper temperature control during transport is crucial to prolong shelf life and ensure that the product retains its appearance. Three ingredients are needed:

- *Chill temperatures*
  Wherever possible fish and shellfish should be held in chill conditions, in transport and in storage.
- *Ice*
  Proper icing is the most effective method of preservation of fin fish. If there are doubts about chill temperatures in transit, use a higher ratio of ice than usual. If problems occur and are likely to prove expensive, equipment is available to monitor product temperature in transit to determine the nature of the problem.
- *Transport*
  Wherever possible use refrigerated vehicles

TRANSPORT OPTIONS

Own transport	Convenient for local deliveries but if using regularly and for longer distances, check that when you have taken time and depreciation into account it actually is cost effective.
Specialist fish operators	A number of Scottish based companies offer good services to a wide variety of European destinations. Use refrigerated services where possible. Some incoming lorries also available.
Other refrigerated transport	Often fish can be taken with other loads or as backhaulage for lorries bringing products into Scotland.
General haulage	A number of companies offer nationwide collection and delivery services to destinations within the UK. Prepared to take small volumes, frequent, flexible and relatively cheap. Cost penalties for small consignments are small. But problem that service not refrigerated. Check on arrival condition of product.
Rail	Disadvantage that no refrigeration but some businesses find cost effective. Check on time in transit and arrival condition.

CHOICES

Choices between the different options should be made by looking at:

- Customers' choice.
- Handling standards: temperature control in transit,
temperature control at depots,
loading and unloading care.

- Reliability in delivery.
- Fit with customer transport system.
- Flexibility.
- Speed.
- Costs.

Talk to people who have used these systems and contact a variety of haulage companies for information. The advice of HIDB/HIE can also be sought because they have a great deal of experience in this area. It is worthwhile doing a bit of research on this because:

- Rates vary considerably between companies.
- Rates vary from different locations.
- Rates vary to different destinations.
- Rates vary with load size.
- Rates vary with frequency of orders.
- Rates vary for different qualities of service.

It is worthwhile checking on the transport arrangements of your neighbours. Possibly by grouping loads more cost effective rates can be obtained.

It is sometimes possible to get very good 'once off' rates from carriers, if your consignment completes a load. If you really want a regular service, check whether this rate would operate permanently or whether you would be better going elsewhere.

However competitive the rates, if your customers are not satisfied with the condition of products they receive, look elsewhere!

AIR FREIGHT

All fresh fish and shellfish for destinations outside Europe goes by air as does a small amount of product to European destinations. Exporters arrange delivery to the freight depot of their airline, preferably by refrigerated vehicle. This must be watched carefully because chill storage at many airports is limited and timing of delivery must be scheduled as close to the departure of flights as is allowed and safe. The freight forwarder will require an hour for documentation after which the airline needs four hours for handling.

Unitised loads are normally used – either closed standard containers or pallets covered with restraining nets. Closed containers give better protection.

The freight forwarder (Chapter 8) will select suitable airlines. Different airlines have different schedules, rates and reputations. Ideally the freight forwarder requires about a week to search for the most cost effective options. Three days are usually the absolute minimum time in which they can manage things.

ACTION POINTS

1 Do your practices and facilities meet legal requirements?

Growing ....
Harvesting ....
Processing ....
Labelling ....
Packing ....

2 Do your current practices and facilities meet the standards required by your customers?

Growing ....
Harvesting ....
Processing ....
Labelling ....
Packing ....

3 Do your practices and facilities meet the standards of your trade association?

Growing ....
Harvesting ....
Processing ....
Labelling ....
Packing ....

4 Do you expect any of these standards and recommendations to change? If so what preparations have you made to meet new requirements?

....

5 Have you checked recently with customers that your product is arriving in satisfactory condition?

Appearance of fish ....
Freshness of fish ....
Grading of fish ....
State of boxes ....
Timing of delivery ....

If the answer is no, what steps have you taken to improve?

....

6 Do you maintain regular contact with a number of carriers to ensure that you are using the most cost effective system for the quality of service you want?

....

10
Trade Associations

Introduction
Marketing and trade associations
Advertising and public relations
Push and pull promotion
Making trade association advertising work
Advertising is expensive!
Quality control
Market research
Other roles
Making the most of trade associations
Useful addresses

INTRODUCTION

Trade associations are very important for small businesses because they make it possible to do things that small businesses they could not otherwise afford but which are very necessary for success. This chapter looks at marketing by trade associations, explains its importance and identifies the responsibilities of both the trade association and its members towards each other.

MARKETING AND TRADE ASSOCIATIONS

Trade associations have an important role in the marketing of the products of members through:

- Advertising and public relations
- Quality control
- Market research

ADVERTISING AND PUBLIC RELATIONS

It makes sense for Scottish fish farmers to promote their products jointly with other similar producers because:

- It is difficult for the individual farmer to develop markets by themselves because they do not have the money.
- It is much easier to develop a general 'Scottish' brand name belonging to all farmers than to create individual brand names: this helps to create customer loyalty.
- By pooling resources the total effect is much greater than each farmer acting alone.

PUSH AND PULL PROMOTION

Individual farmers *push* the product into the marketing channel – they talk to trade customers to persuade them to buy. In most cases it would be quite impossible to advertise to final consumers because the money needed is far too great for the individual farmer to afford.

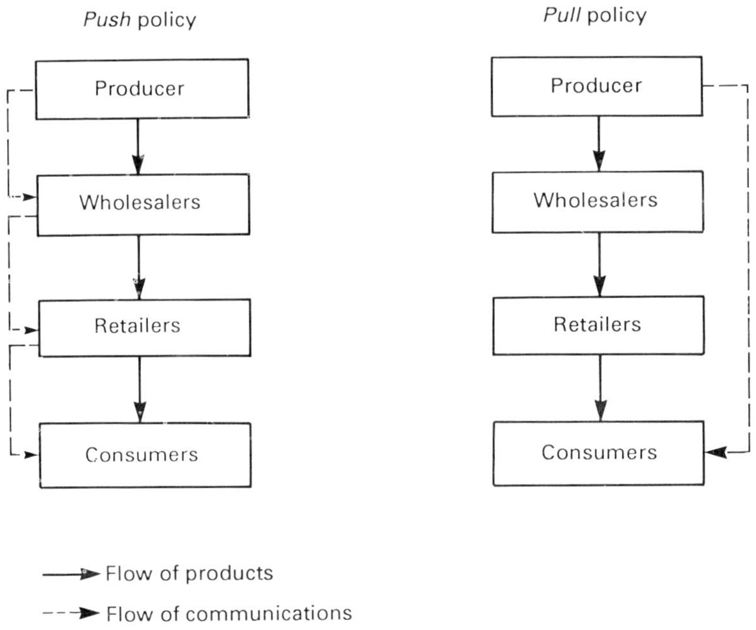

Fig. 10.1

But someone has also to *pull* the product through from the other end, by persuading final consumers that they should buy. Advertising and public relations work directed at the general public can:

- Make people aware of the product.
- Create interest.
- Make people want to buy.
- Persuade existing customers to buy more.

It is not sensible to try to do all of these at once, but depending on the needs of the market, all are reasonable targets at different times for the trade associations. It is worth remembering that:

- There are thousands of products in the typical supermarket.
- There are many different types of fish and shellfish in the fishmongers.
- Most restaurants with farmed fish also offer a choice of other dishes.
- People actually do not spend a lot of time thinking about food, but they do respond when particular products are brought to their attention.
- Other products are advertised to the general public and your product will lose out if it is not.

This does not mean that trade associations should only advertise to final consumers. There are also benefits in advertising jointly to members of the trade. It does mean that if the association can afford it this can be very valuable.

MAKING TRADE ASSOCIATION ADVERTISING WORK

There have been very successful advertising and public relations campaigns by trade associations in the food industry, including those of fish farmers trade associations. But there have also been examples of costly failures. Usually where this has happened it has been due to one of the following:

- *Budgets which were too small*
  Advertising is expensive as you can see below but fish farmers have to compete on a similar level with other products if they are to have any effect. Without adequate funds the trade association cannot work for you.
- *Insufficient commitment by the association and its members*
  If you do not support actions and are not willing to make an investment in advertising and see it through, it will not be successful. It takes time to make an impact – don't be impatient!
- *Unclear objectives*
  It is important that there is a consensus so that the association receives the support of its members but at the same time members must be willing to let the experts take the decisions.

ADVERTISING IS EXPENSIVE!

Expenditure on advertising 1988!

	(£ million)
Coffee	55
Milk	18
Beer	92
Sea Fish Industry Authority	3.4
Bird's Eye Fish Advertising	3.6
Ross Youngs Fish Advertising	2.3
Scottish Salmon Board	1.2

QUALITY CONTROL

There is no point in spending the scarce money of members on advertising if products do not match the advertising image. If the quality standards of members are variable, expenditure on advertising is wasted. Poor quality can only harm the general image of the industry – something which is very serious when you are competing with fish farming industries from other countries.

Quality control schemes can be of different types but the main distinction is between voluntary and controlled schemes:

- *Voluntary*
  A code of conduct for farming, harvesting and post harvest handling recommended to members.
- *Controlled*
  A code of conduct for farming, harvesting and post harvest handling which is compulsory for those who agree to take part. All fish boxes are numbered and penalties can be imposed for failure to comply. These schemes are normally backed by random inspection of plant and products.

Controlled schemes are more expensive to run but quality standards should be higher. Often for a new association the best first step is to start with a voluntary scheme and later with more experience and commitment to move on to a controlled system.

Details of the schemes currently in operation can be obtained from your trade associations.

MARKET RESEARCH

All members are likely to have similar information needs:

- Up-to-date information on prices.
- Information on current market conditions.

- New customer leads.
- Forecasts of future conditions.
- Market studies.

It is cheaper to have this information provided by the trade association than to do it yourself.

OTHER ROLES

Of course the trade association is not only concerned with marketing. It plays a valuable role also in *Representation* and *Research*.

There are many pressures on fish farmers from government at various levels and from interest groups. The industry must see that, where sensible, policies and attitudes towards it are favourable. Some examples where this has been necessary in recent years are:

Local governments
- Facilities
- Planning permissions

UK Government
- Size of levies
- Use of water
- Regulations on water quality

European Community
- Trading conditions and tariffs
- Planning and resource use
- Environmental controls
- Hygiene regulations in processing
- Disease prevention
- Grants

The best way of getting the policies you want is through the trade association because it can:

- Pool resources to deal with common problems.
- Employ specialists.
- Be recognised as the 'voice of the industry'.

The same is true of research. By combining the budgets of individual farmers it is possible to commission valuable research to help make production more efficient and therefore profitable.

MAKING THE MOST OF TRADE ASSOCIATIONS

It is always more of a challenge to do things as a group than individually and trade associations can have certain problems:

- *Free riders*
  Farmers who reap the benefits from promoting the products of the industry but who will not pay the levy.
- *Poor communications*
  These can create tension between the association and the membership leading to a lack of commitment on both sides.

These problems can be avoided by:

Trade association

- Carrying out the management of the association efficiently and keeping members well informed.
- Having sufficient budgets to do this.
- Making sure that they have the support of members for major policy decisions by providing clear explanations and opportunities for discussion.

The industry

Supporting the association by:

- Membership.
- Paying levies promptly.
- Responding to initiatives.
- Providing information.
- Joining in discussions – you can't criticise if you don't take part!
- Listening to the advice of experts.

In this way the associations representing Scottish fish farmers will continue to contribute to the prosperity of the industry.

CHECKLIST

1 Are you a member of your trade association?

....

2 How informed are you, about the marketing plans of your trade association?

Very well informed ....
Well informed ....
Badly informed ....

3 Do you read the information sent to you?

Always ....
Sometimes ....
Never ....

4 Do you attend meetings?

Usually ....
Sometimes ....
Never ....

5 List three ways in which you have helped your trade association during the past year

(1) ....
(2) ....
(3) ....

USEFUL ADDRESSES:

Association of Scottish Shellfish Growers
Old Ferryhouse
Glenelg
By Kyle
Ross-shire IV40 8JZ
Tel: 059 982 253

British Trout Association
PO Box 189
London SW6 7UT
Tel: 071 385 1159

Scottish Salmon Growers Association
Drummond House
Scott Street
Perth PH1 5EJ
Tel: 0738 35420

Shetland Salmon Farmers Association
80 Commercial Street
Lerwick
Shetland ZE1 0DL
Tel: 0595 5579

Shellfish Association of Great Britain
Fishmongers' Hall
London Bridge
London EC4R 9EL
Tel: 071 283 8305

Scottish Salmon Smokers' Association
PO Box 118
Edinburgh EH3 7AS
Tel: 031 220 2256

See also Useful Addresses section at the end of the book for further information.

Useful Addresses

Association of Scottish Shellfish Growers
Old Ferryhouse
Glenelg
By Kyle
Ross-shire IV40 8JZ
Tel: 059 982 253

Bank of Scotland
International Division
8 York Place
Edinburgh EH1 3EP
Tel: 031 243 5089

British Food Export Council
301–344 Market Towers
New Covent Garden Market
London SW8 5NQ
Tel: 071 622 0188

British Trout Association
PO Box 189
London SW6 7UT
Tel: 071 385 1159

Clydesdale Bank PLC
International Services
150 Buchanan Street
Glasgow G1 2HL
Tel: 041 248 7070

Department of Agriculture and Fisheries for Scotland
Marine Laboratory
PO Box 101
Victoria Road
Aberdeen AB9 8DB
Tel: 0224 876544

Dun and Bradstreet
Business Information Services
11th Floor
Centre City Tower
Birmingham
Tel: 021 631 2323

Euroinfocentre
Bothwell Street
Glasgow
Tel: 041 221 0999

Department of Trade and Industry
Victoria Street
London SW1
Tel: 071 215 7877

Food From Britain
301–344 Market Towers
New Convent Garden Market
London SW8 5NQ
Tel: 071 720 2144

Highlands and Islands Development Board (HIDB)/Highlands and Islands Enterprise (HIE)
Bridge House
20 Bridge Street
Inverness IV1 1QR
Tel: 0463 234171

Institute of Freight Forwarders
Suffield House
9 Paradise Road
Richmond
Surrey TW9 1SA
Tel: 081 844 2266

NatWest Bank
International Office
14 Blythswood Square
Glasgow G2 6AT
Tel: 041 221 1838

Royal Bank of Scotland PLC
International Division
PO Box 106
Moncrieff House
69 West Nile Street
Glasgow G1 2LT
Tel: 041 248 2777

Scottish Agricultural Organisation Society (SAOS)
18/19 Claremont Crescent
Edinburgh
EH7 4JW
Tel: 031 556 6574

The Scottish Business Information Service
Scottish Science Library
National Library of Scotland
33 Salisbury Place
Edinburgh
EH9 1SL

Scottish Export Office
Industry Department for Scotland
Alhambra House
45 Waterloo Street
Glasgow G2 6AT
Tel: 041 242 5493

Scottish Salmon Growers Association
Drummond House
Scott Street
Perth PH1 5EJ
Tel: 0738 35420

Scottish Salmon Smokers' Association
PO Box 118
Edinburgh EH3 7AS
Tel: 031 220 2256

Sea Fish Industry Authority
Sea Fisheries House
10 Young Street
Edinburgh
EH2 4JQ
Tel: 031 225 2515

Shellfish Association of Great Britain
Fishmongers' Hall
London Bridge
London EC4R 9EL
Tel: 071 283 8305

Shetland Salmon Farmers Association
80 Commercial Street
Lerwick
Shetland ZE1 0DL
Tel: 0595 5579

Index

Books published by
Fishing News Books

Free catalogue available on request from Fishing News Books, Blackwell Scientific Publications Ltd, Osney Mead, Oxford OX2 0EL, England